解析西格蒙德·弗洛伊德
《梦的解析》

AN ANALYSIS OF
SIGMUND FREUD'S
THE INTERPRETATION OF DREAMS

William J. Jenkins ◎ 著

张和龙 ◎ 译

目 录

引言 …………………………………………… 1
   西格蒙德·弗洛伊德其人   2
   《梦的解析》的主要内容   3
   《梦的解析》的学术价值   4

第一部分：学术渊源 ………………………… 7
   1. 作者生平与历史背景   8
   2. 学术背景   13
   3. 主导命题   17
   4. 作者贡献   21

第二部分：学术思想 ………………………… 25
   5. 思想主脉   26
   6. 思想支脉   30
   7. 历史成就   35
   8. 著作地位   39

第三部分：学术影响 ………………………… 45
   9. 最初反响   46
   10. 后续争议   51
   11. 当代印迹   55
   12. 未来展望   59

术语表 ………………………………………… 63
人名表 ………………………………………… 68

CONTENTS

WAYS IN TO THE TEXT 73
    Who Was Sigmund Freud?   74
    What Does *The Interpretation of Dreams* Say?   75
    Why Does *The Interpretation of Dreams* Matter?   77

SECTION 1: INFLUENCES 79
    Module 1: The Author and the Historical Context   80
    Module 2: Academic Context   85
    Module 3: The Problem   90
    Module 4: The Author's Contribution   94

SECTION 2: IDEAS 99
    Module 5: Main Ideas   100
    Module 6: Secondary Ideas   105
    Module 7: Achievement   111
    Module 8: Place in the Author's Work   116

SECTION 3: IMPACT 121
    Module 9: The First Responses   122
    Module 10: The Evolving Debate   127
    Module 11: Impact and Influence Today   132
    Module 12: Where Next?   137

Glossary of Terms   142
People Mentioned in the Text   147
Works Cited   149

引 言

要 点

- 西格蒙德·弗洛伊德是奥地利神经病学家*（专门治疗神经系统与大脑疾病的专家），精神分析学*（关于无意识心理*的治疗与理论方法）的创始人。
- 在《梦的解析》中，弗洛伊德认为通过分析和解读梦的意义，可以理解无意识心理中的各种欲望。
- 学术界将《梦的解析》视作弗洛伊德最重要的著作。该书对心理学*（研究人类心理和行为的科学）和精神病学*（精神障碍的治疗与研究）产生了很大的影响。

西格蒙德·弗洛伊德其人

西格蒙德·弗洛伊德出生于1856年，一生大多数时间生活在奥地利维也纳。1881年，他获得维也纳大学医学学位后，在当地一家地方医院的精神病科工作。这份工作让他亲身接触到各种精神病人，获得了临床工作的实际经验。

1885年，弗洛伊德跟随著名法国神经学家让-马丁·沙可*进行研究。与沙可合作期间，弗洛伊德意识到许多精神疾病可以追溯到病人早年的心理创伤事件。这些创伤事件储存在无意识心理中。1893年，奥地利医生约瑟夫·布洛伊尔*告诉弗洛伊德：他手头有一位病人出现了一些极不寻常的临床症状，但是找不到明显的临床病因。布洛伊尔医生没有对病人进行药物治疗，而是采用了心理疗法——让病人倾诉个人经历。一段时间后，布洛伊尔医生发现"倾诉疗法"改善了病人的临床症状。1895年，弗洛伊德和布洛伊尔在合写《癔症研究》中记述了这种心理治疗的新方法。

与沙可、布洛伊尔合作的经历激发了弗洛伊德的思想灵感，即无意识心理对人类行为的决定性作用。他开始将梦视作无意识心理表达其欲望的一个途径。1899 年，弗洛伊德的经典名作《梦的解析》出版后广为流传，由此开创了心理学与精神病学的精神分析视角。这本著作自此在世界各地出版，一个多世纪后仍然不断重印。

《梦的解析》的主要内容

在弗洛伊德之前，大多数人相信梦没有目的，也没有意义。弗洛伊德对此提出了激进的观点：梦作为无意识心理的产物，实际上隐含着丰富的意义。

在《梦的解析》中，弗洛伊德认为梦隐含着做梦者未实现的无意识欲望等富有意义的信息。在他看来，每一个梦都是一份伪装过的信息。要想解读出其意义，必须进行梦的解析*。通过联想而进行交谈的过程中，做梦者可能会将梦所隐含的信息保存下来。

弗洛伊德在开篇综述中回顾了已有的梦学理论文献。他讨论了古代人将梦看成是神灵或魔鬼之作的观点，也审视了同时代的其他观点，即梦是大脑感观区域清醒状态或神经活动*（细胞层面的电子和化学作用过程）的延伸。

弗洛伊德演示了析梦、释梦的方法。他描述了如何让病人展开自由联想*，然后考量梦中的每一个细节。在自由联想中，一方（比如说治疗师）先说一个词，另一方（病人）立刻用另一个词进行回应。这些词语也许没有明显的联系，但是治疗师可以通过排序来解读病人无意识心理的运作机制。

弗洛伊德还表明，所有的梦都代表了无意识欲望的实现。我们的意识通过某种方式审查这些欲望，因此它们只能在我们的梦中显现。梦中的故事，或"显梦内容"*，掩盖了梦的真实意义，或"隐梦"内容*。弗洛伊德描述了大脑将欲望转换为梦境的几种方法。这些方法有助于读者更好地解析梦境。

如果欲望令人不适，处于未审查状态，得不到承认，无意识就将它们转化为梦。在弗洛伊德看来，这些欲望大多是性冲动。这些性冲动往往在幼年时孩子与父母互动过程中形成。弗洛伊德相信，孩子将性冲动和暴力冲动的对象指向父母，是正常的人类成长过程。在他看来，健康者与精神病患者的差别在于后者放大了这些正常的冲动。

弗洛伊德在描述他的笼统的心理理论*时，借用了反射循环论*——心理接受感官信息、身体则以某种方式作出回应的一种简单联系。这种对心理概念的生物学理解在当时比较普遍，反映了弗洛伊德作为一名临床医生的科班背景。他认为心理是由两部分组成：无意识和意识。无意识没有直接通往意识的通道，主要依赖梦作为途径，把无意识的需要和欲望传送到意识域*。这一思想是弗洛伊德后期很多著作的基石，这也让他在同行之中显得卓然不同。

《梦的解析》的学术价值

学界普遍认为，《梦的解析》是弗洛伊德最重要的代表作。在该书中，他奠定了未来很多思想的基础。该书使他在心理学和精神病学领域开创了精神分析学派。该书还为精神疾病的治疗提供了方法。精神分析学派注重无意识心理在推动人类意识行为中的作用。弗洛伊德还提出了性心理成长的"阶段说"*，即使用精神分析学的原则来解释（非常粗略）人的正常成长，其中涉及性快感来源的不同阶段。弗洛伊德相信，成年人的人格*特征反映了人在经过不同发展阶段后性冲动的消除。

《梦的解析》与弗洛伊德的精神分析视角在心理学与精神病学这两门学科的发展过程中起到了巨大作用。这部著作帮助人们使用全新的方法思索无意识心理（例如该书认为梦含有来自无意识的有意义的信息）。弗洛伊德的思想激发了众多的追随者，精神分析学的影响也越来越大。有些人在这部著作的启迪下，对无意识心理的本质特性和性本能的重要性提出了不同的观点。作为这些争议的结果，有些追随者建立了他们自己的理论视角。这些人在理念上明显是以弗洛伊德的原创思想为基石，因而被称作"新弗洛伊德派"*。

这部著作引发的种种批评可能与著作一样举足轻重。几乎从一开始，弗洛伊德的一些同代人指责他的思想存在错误。批评者们认为，他的学说带有公开的八卦性（所举证的例子可能并不具有真正的代表性），因为这些思想来自他与病人的合作以及他本人的自我分析。除了八卦性外，这些思想还无法得到验证。科学要求思想学说应该能够被检验、被验证——但无人能观测到无意识心理。尽管这本著作似乎很不科学，但是弗洛伊德明显是个珍视科学的人。阅读这本著作，这一点清晰明了。然而，科学的进步打开了一道用新方法探索旧思想的大门，使我们能更好地去验证弗洛伊德从他本人和病人的经历中所得出的结论。

第一部分：学术渊源

1 作者生平与历史背景

要点

- 《梦的解析》为弗洛伊德的理论奠定了基础。他的理论影响巨大，彻底改变了心理学（研究人类心理与行为的科学）。
- 弗洛伊德对他本人的梦和父亲之死的分析成为《梦的解析》的基础。
- 维多利亚时代*（跨越19世纪与20世纪初的历史时期，以英国王室维多利亚女王在位为标志）的道德观影响了弗洛伊德的大半生，对他的理论的形成产生了很大的影响。

为何要读这部著作？

《梦的解析》是弗洛伊德最负盛名的著作。该书出版于1899年，弗洛伊德从此声名鹊起，被誉为"精神分析学理论与实践之父"。精神分析学运动主导了心理学研究领域数十年。[1] 尽管他的许多原创思想被证明是有问题的，但是毋庸置疑，弗洛伊德是心理学历史上最有影响力的人物之一。正如一位学者所言，"弗洛伊德与[政治哲学家]卡尔·马克思*、[进化论的奠基者]查尔斯·达尔文*、[物理学家]阿尔伯特·爱因斯坦*一起跻身天才之列。这些天才对于我们认识我们所在的世界、我们在世界中的位置，以及在认识方式上，都烙下了极其重要的印记。"[2] 弗洛伊德的影响如此之大，"甚至我们的语言……都浸透了'弗洛伊德式'的意义，言语失误*、压抑*、投射*、合理化*、防御*等等"——所有来自弗洛伊德理论的术语都与无意识言语、无意识心理和行为，以及心理

治疗过程息息相关。³

奥地利精神病学家*A. A. 布里尔*于1913年出版了这本著作的第一个英译本。他将《梦的解析》誉为"弗洛伊德最伟大、最重要的著作"⁴。在这部著作中，弗洛伊德奠定了人类心理与人格结构的理论基础。在这部著作的指引下，一些医学专业人士追随弗洛伊德的脚步，成为精神分析学家。该书是精神分析学与心理学历史上至为重要的代表性文本，可以让读者洞悉弗洛伊德的主要思想，即梦在传导无意识心理的信息过程中扮演了重要角色。

> "梦是反常心理结构*链条中的第一个连接点，而临床医生们出于医疗实践的原因，对这个链条中的其他连接点，如歇斯底里恐惧症*、固恋*和幻觉*，势必产生兴趣。"
>
> ——西格蒙德·弗洛伊德：《梦的解析》

作者生平

西格蒙德·弗洛伊德出生于1856年，一生大部分时间生活在他的家乡奥地利维也纳。他在维也纳大学攻读医学专业，于1881年获得医学学位。第二年，他在维也纳综合医院任职诊所助理，在神经病理学家*西奥多·梅涅特*领导下的精神病科工作（神经病理学是对人的神经系统障碍与大脑的研究）。1885年，弗洛伊德获得基金资助，跟随"神经病学之父"、法国人让-马丁·沙可从事研究。沙可向弗洛伊德讲授歇斯底里症（一种神经官能症，其特点是极端狂躁）的临床症状，以及如何运用催眠术*进行治疗。⁵

弗洛伊德回到维也纳后，开了一家私人诊所，专门治疗患有神经系统疾病的病人。1893年，他开始与奥地利医生约瑟夫·布洛伊尔协同工作。两人合作撰写了《癔症研究》[6]（1895），书中提出将自由联想法（心理分析师给病人提供一系列词语，让病人对这些词语立即作出回应）作为治疗歇斯底里症的方法。在这个时期，弗洛伊德开始对他本人的梦境进行分析和阐释。他在《梦的解析》中记载了这一经历和他与病人合作的经历，还有他父亲去世一事。在该书第二版的前言中，弗洛伊德称，它是"我的自我分析的一部分，是对父亲去世的反应——换言之，是对一个人生命中最重要事件、最惨痛损失的反应。"[7]

《梦的解析》第一版于1899年出版后，弗洛伊德创建了后来闻名遐迩的维也纳精神分析学会。他在自己的职业生涯中还出版了多本著作，继续阐发他在《梦的解析》中所提出的观点。作为犹太人，弗洛伊德受到了来自邻国纳粹德国*反犹主义*思想的影响（反犹主义是指对犹太人的仇视）。1938年，德国吞并了奥地利。几个月后，弗洛伊德和他的家人逃到了英国。他于1939年9月23日在伦敦逝世。[8]

创作背景

瑞士精神病专家卡尔·荣格*曾跟随弗洛伊德进行研究。他说："弗洛伊德所身处的历史环境是弗洛伊德现象出现的基础与必要条件。"[9]确如所言，弗洛伊德所生活的时代在他的思想的形成过程中起到了很大的作用。他所受到的最深刻的影响之一是19世纪在欧洲风行一时的维多利亚道德观。维多利亚人展示了"对性的强烈道德关注……性不仅需要由个人道德或者由保持警惕的家庭来管控，而且因为性可能会影响到所有的人，所以也成为政治与社会

问题。"¹⁰

在这个时期,整个社会还对"科学唯物论*〔认为一切心理活动都是大脑的功能〕和理性主义*〔以逻辑为主导倾向的哲学态度〕"产生越来越浓厚的兴趣。荣格认为,"这是弗洛伊德得以成长的母体,而正是这个母体的心理特征在预先设定的轨道上形塑了他。"¹¹

弗洛伊德相信,无意识心理影响我们的行为。我们可以通过分析梦境来管窥无意识中没有实现的欲望。弗洛伊德认为,无意识心理在压抑性态度与性冲动中起到了核心作用。考虑到维多利亚时代性压抑的背景,我们不应该对此感到惊讶。

---

1. B. M. 索恩、T. B. 亨里:《心理学史与心理学体系的联系》,波士顿:霍顿·米夫林出版公司,2005 年。
2. 罗伯特·S. 沃勒斯坦:"弗洛伊德的精神分析学在 21 世纪的意义:它的科学与研究",《精神分析心理学》第 23 卷,2006 年第 2 期,第 302—326 页。
3. 沃勒斯坦:"弗洛伊德的精神分析学在 21 世纪的意义",第 303 页。
4. 西格蒙德·弗洛伊德:《梦的解析》,A. A. 布里尔译,丹尼尔·T. 奥哈拉与吉娜·马苏奇·麦肯济导论和注释,纽约:巴恩斯和诺贝尔图书出版公司,2005 年,第 9 页。
5. 弗洛伊德:《梦的解析》,第 xi—xii 页。
6. 西格蒙德·弗洛伊德和约瑟夫·布洛伊尔:《癔症研究》,詹姆斯·斯特拉齐译,伦敦:霍加斯出版社,1955 年。
7. 弗洛伊德:《梦的解析》,第 6 页。
8. 弗洛伊德:《梦的解析》,第 xiii—xix 页。
9. C. G. 荣格:"西格蒙德·弗洛伊德写作时的历史语境",《人格学刊》第 1 卷,

1932年第1期，第48—55页。

10. "西格蒙德·弗洛伊德写作时的历史语境"，《哥伦比亚学院核心课程》，登录日期2016年1月2日，http://www.college.columbia.edu/core/content/writings-sigmund-freud/context。

11. 荣格："西格蒙德·弗洛伊德写作时的历史语境"，第49页。

2 学术背景

要点

- 许多人将西格蒙德·弗洛伊德视作精神分析学的理论与实践之父,而实际上,在弗洛伊德出生之前 80 年,德国医生和催眠师弗朗兹·安东·麦斯麦尔*已经奠定了精神分析学的基石。
- 在弗洛伊德进入这一领域之前,神经精神病学*(探讨神经系统在精神疾病中的作用)在论述心理疾病时主要采用生理学分析视角。
- 弗洛伊德与让-马丁·沙可、约瑟夫·布洛伊尔两位临床医生的合作,对他创建精神分析学产生了很大的影响。

著作语境

很多人将《梦的解析》视作西格蒙德·弗洛伊德最重要的作品。它将精神分析学,有时也称作精神动力治疗学*(强调无意识动能对人类行为产生影响的思想流派)看作是心理学与精神病学的主导学说。

弗洛伊德坚信,无意识心理深刻地影响着我们有意识的行为和人格。这一崭新的学说为治疗心理疾病的诸多学科带来了根本性的改变。的确,这些改变不断撼动着人类文化的基石。不过,弗洛伊德并不是首先提出无意识心理这一概念的人。有人认为,1775 年"弗朗兹·安东·麦斯麦尔医生与驱魔师约翰·约瑟夫·加斯纳*1之间的冲突"奠定了精神分析和无意识理论的基础。("驱魔"是指从患病的躯体中祛除诸如魔鬼等邪恶力量的仪式。)

加斯纳，一位深受欢迎的治疗师，通过使用宗教方法来消除人体的各种疾病。1775年，当著名的社会与文化运动，即启蒙运动*所提倡的理性在欧洲逐渐成为主流时，加斯纳与他的神秘主义疗法受到了质疑。麦斯麦尔提供了一种替代性的方法。作为一个训练有素的医生，他采用更加科学的方法来治疗精神疾病。但医疗机构最终拒绝了麦斯麦尔的设想。不过，他对通灵治疗方法的反对带来了一场变化，专业医疗人士最终接受了精神分析学治疗方法。[2]

> "弗洛伊德开启了全新动力学学派的新纪元，这个学派用正统的思想、严格的组织、专业化的期刊、密闭的会员制，以及冗长的入会仪式来管制它的成员。"
> ——亨利·F.艾伦伯格：《无意识的发现：
> 动力精神病学的历史与演化》

学科概览

西格蒙德·弗洛伊德进入这个领域时，精神疾病的治疗方法已经发生了一些改变。此前，人们相信精神病患者乃魔鬼附体，所以加斯纳这样的驱魔师使用通灵手法进行治疗。不过，到了19世纪晚期，这一领域中的大多数专业人士采取更加系统的生物学手段来理解这些疾病。他们聚焦人体有形的（也就是说具体的）、可以直接观察到的方面，比如说他们的身体状态。科学研究在当时发生了彻底的转向，以致弗洛伊德不得不这样表述：他通过梦的解析来聚焦无意识，并未"逾越神经病理学研究的界限"[3]（"神经病理学"此处是指对神经系统与大脑疾病的研究）。他在提到恐惧症*（非理性害怕）和妄想（错误的幻想）时又补充说道："如果不能解释梦境的起源，就去理解各种恐惧症、固恋与妄想的念头，及其疗法的重要性，这样的努力是徒劳无益的。"[4] 显然，弗洛伊德担心的是，主流学界在很大程度上忽视了心理因素对人类行为的影响。

不过，该领域中的一些人与弗洛伊德不谋而合。他们相信，无意识心理与精神因素都会对精神病人的症状产生影响。弗洛伊德在理论草创阶段与沙可等同行的合作，帮助他形成了《梦的解析》中的思想。

学术渊源

基金的支持让弗洛伊德有机会前往巴黎跟随法国医生让-马丁·沙可从事研究。沙可被很多人视作神经病学之父。他专门医治那些症状奇特且没有明显身体致病原因的患者。经过漫长的职业生涯，沙可最终相信，他的病人罹患"歇斯底里症的病因是患者对过去创伤事件所作出的强烈情绪反应"[5]。在与沙可的合作中，弗洛伊德得出结论认为，神经官能症*（涉及焦虑的精神疾病）源于无意识，并导致了这些奇特症状。

弗洛伊德与德高望重的奥地利医生约瑟夫·布洛伊尔的合作也对他产生了深刻的影响。布洛伊尔帮助弗洛伊德完成了他的医疗实践，在很多方面成为他的研究导师。布洛伊尔向弗洛伊德讲述了他收治到的最有趣的一位女病人，后来她以安娜·欧*闻名于世。他们后来在1895年合著的《癔症研究》[6]中记述了这个病人的故事。安娜的奇特病症主要有不明原因的咳嗽、半身瘫痪，甚至会产生栩栩如生的幻觉。布洛伊尔最终将她诊治为歇斯底里症。然而他发现，当他与安娜讨论这些幻觉时，她的症状慢慢消失了。这一交谈疗法激发了弗洛伊德的灵感。他将创伤导致的歇斯底症思想与布洛伊尔的交谈疗法融为一体，开启了他独树一帜的理论方法。在弗洛伊德之后，自由联想法（在治疗过程中，要求患者对语词作出即时性回应而不需要进行自我审查）与梦的解析法（发现梦境所传达的无意识心理信息的过程）已经成为精神分析学的奠基石。[7]

1. 亨利·F.艾伦伯格：《无意识的发现：动力精神病学的历史与演化》，纽约：基础图书公司，2008年，第53页。
2. 艾伦伯格：《无意识的发现：动力精神病学的历史与演化》，第53—57页。
3. 西格蒙德·弗洛伊德：《梦的解析》，A. A. 布里尔译，丹尼尔·T. 奥哈拉与吉娜·马苏奇·麦肯济导论和注释，纽约：巴恩斯和诺贝尔图书出版公司，2005年，第9页。
4. 弗洛伊德：《梦的解析》，第3页。
5. 理查德·韦伯斯特："弗洛伊德、沙可与歇斯底里症：消失在迷宫中"，登录日期2016年1月3日，http://www.richardwebster.net/freudandcharcot.html。
6. 西格蒙德·弗洛伊德和约瑟夫·布洛伊尔：《癔症研究》，詹姆斯·斯特拉齐译，伦敦：霍加斯出版社，1955年。
7. 韦伯斯特："弗洛伊德、沙可与歇斯底里症：消失在迷宫中"。

3. 主导命题

要点

- 弗洛伊德与他的同代人想要更好地理解某些精神疾病的心理原因。
- 弗洛伊德的大多数同行试图从生理学层面来解释精神疾病的症状。因此,他们很大程度上忽视了无意识心理与梦的解析。
- 弗洛伊德相信无意识心理在很多精神疾病的发病中扮演了重要角色。他还认为梦能够让无意识心理中未实现的欲望呈现出来。

核心问题

在《梦的解析》中,西格蒙德·弗洛伊德提出了他的观点,认为梦有助于深入了解人的无意识心理,解析梦有助于治疗精神疾病。弗洛伊德撰写这本著作的时候,主流医学界极少关注梦与无意识心理。他们致力于从身体层面来确定精神疾病的病因。在当时,描述神经系统的技术手段寥寥无几。相反,医疗专家们认为人的行为是反射循环的结果:循环建立在人脑基础之上,涉及到感官输入*,即通过视觉、听觉等器官输入,和能动输出*,即身体作出反应。弗洛伊德注意到,"反射弧*始终是考量每一个心理活动的模型"[1],因此他在《梦的解析》中所提出的心理理论沿袭了这一基本的结构。

在弗洛伊德看来,当时的核心问题在于确定无意识心理在影响有意识的行为中起到了什么样的作用,如果有作用的话。在

《梦的解析》中，弗洛伊德认为对无意识的强调是正确的，指出其他临床治疗人员可以使用特定步骤来分析和阐释梦境。弗洛伊德使用科学语言来表达这些思想，希望这本著作能够吸引来自科学界与通灵界的众多同行。

> "不可否认的是，弗洛伊德的反射能动理论偏离了19世纪的主流思想。"
>
> ——尼玛·巴西里："弗洛伊德与脑物质"

参与者

弗洛伊德确信，某些类型的精神疾病实质上源自心理层面，而非生理层面。他与法国神经病学家让-马丁·沙可、奥地利医生约瑟夫·布洛伊尔的合作强化了他的信念。然而，当时很多人反对他的观点。弗洛伊德受过专业的医学训练，越来越专注于对心理的科学理解，提出了精神分析学的理论。但弗洛伊德秉持心理疾病可能会引发身体病症的观点，从而"与正统医学界公开决裂"[2]。一些批评者将梦看作是对"引发睡眠紊乱的简单的刺激反应"[3]。有鉴于此，主流医学界认为梦没有任何目的。因为没有目的，梦对理解人类心理和歇斯底里症的症状不会提供任何特别帮助。

然而，弗洛伊德相信，"梦是有意义的，尽管其意义是隐晦的；梦是某种其他思想过程的替代品，只有正确分析这个替代过程，才能揭示出梦的隐秘内涵。"[4]他撰写《梦的解析》一书，向他的医学界同行传达他的观点。他相信，总有一天，当医生无法确定生理或者身体层面的反常病症时，医学界就会使用他的方法来治疗饱受身体病症痛苦的病人。

当代论战

弗洛伊德撰写《梦的解析》时，反射行为理论颇为流行。弗洛伊德本人在撰写这本著作前，也从反射的层面描述人的行为。正如一位学者所说，"弗洛伊德用反射行为来界定某种趋势，即惯性*。"[5] 弗洛伊德使用的"惯性"一词含有特殊的意义；"惯性"显示了"神经系统的趋向——因此，也是心理*消除过度兴奋后的趋向。"[6] 弗洛伊德与他的同代人相信，人在这种过度兴奋的消除过程中获得愉悦。然而，如果让这种兴奋状态不断累积的话，那么就会导致反向背离——回避某事的欲望。

当弗洛伊德使用当时通行的模式提出自己的思想时，他的反射观与更加著名且普遍流行的主流思想实际上存在十分重要的差异。具体来说，弗洛伊德聚焦无意识心理，与当时被称之为"反射论"——感官刺激—行为反应体系——几乎没有相似之处。批评者们注意到了这些差异，攻击弗洛伊德的心理学和梦可以帮助人们理解心理机制的思想。受到他人批评后，弗洛伊德感觉自己没能得到精神病*专业群体的重视和赏识。在该书第二版前言中，他甚至说，"科学界的批评之举只能让我感到这本书注定湮没，被人遗忘。"[7]

---

1. 西格蒙德·弗洛伊德：《梦的解析》，A. A. 布里尔译，丹尼尔·T. 奥哈拉与吉娜·马苏奇·麦肯济导论和注释，纽约：巴恩斯和诺贝尔图书出版公司，2005

年，第425页。
2. 亨利·F.艾伦伯格:《无意识的发现：动力精神病学的历史与演化》，纽约：基础图书公司，2008年，第418页。
3. 弗洛伊德:《梦的解析》，第73页。
4. 弗洛伊德:《梦的解析》，第89页。
5. 尼玛·巴西里:"弗洛伊德与脑物质：论神经精神分析学的重组"，《批判性探索》第40卷，2013年第1期，第83—108页。
6. 巴西里:"弗洛伊德与脑物质"，第91页。
7. 弗洛伊德:《梦的解析》，第5页。

4 作者贡献

要点 ⚷

- 弗洛伊德相信心理或精神积极构筑梦境，传达了关于睡梦者未实现愿望或欲望的信息。
- 弗洛伊德关注无意识心理，为精神疾病的生理病因提供了另一种开创性的解释。
- 在《梦的解析》中，弗洛伊德创造性地整合和拓展了他与让-马丁·沙可、约瑟夫·布洛伊尔两位医生的合作成果。

作者目标

西格蒙德·弗洛伊德撰写《梦的解析》，旨在让读者相信梦有十分重要的意义，对梦进行解析可以洞悉无意识心理中的愿望和欲望。洞悉了无意识心理，就可以用来治疗精神疾病。弗洛伊德在该书第一章用全部内容探讨了当时与梦有关的知识。他总结道，"尽管几千年来人们不断努力，但是在梦的科学阐释方面几乎没有取得进展。"[1]

尽管如此，弗洛伊德纵览了自古以来各种各样关于梦的思想。他提到其中一些梦思想时只是一笔带过——如古代人认为梦是鬼神活动的产物。他将当时比较现代的梦思想划分为几个大类：

- 把梦与清醒状态联系起来的思想
- 关于梦与记忆的思想
- 关于梦是睡眠紊乱或是对刺激作出反应的思想
- 解释我们为什么醒后很容易忘掉梦的思想

- 关于梦与清醒状态之间的心理功能存在差异的思想
- 梦的道德内涵
- 梦可能具有功能的思想
- 精神疾病与梦之间的关系 2

弗洛伊德用同样篇幅的内容来强调梦的重要性，以及梦对我们洞悉人类心理的意义，并以具体的梦作为样本进行解析。他在书中指出，他在与病人的合作中认识到，"梦可能与一连串精神事件相关联，而精神事件的源头可以追溯到储存在记忆中的某个病态 * 欲念。其次是将梦当做某种症状，并运用释梦的方法对此类症状进行阐释。" 3

> "我将在以下文字中证明，有一种心理方法可以用来对梦进行解析，一旦使用了这个解析方法，每一个梦都会呈现出一个有意义的精神结构，这个结构可以在清醒状态时的内心活动中找到它相应的位置。"
>
> ——西格蒙德·弗洛伊德：《梦的解析》

研究方法

一股知识和社会潮流自 17 世纪以来席卷欧洲，它被称为启蒙运动。该运动的核心是强调在看待一切事物时理性与逻辑的重要性。对精神疾病的理解和治疗也是如此。弗洛伊德时代的大多数临床医生关注导致精神疾病的生理因素。作为医生，弗洛伊德受到的也是这种传统训练。然而他感到困惑不解的是，大量的病人出现了奇特的身体病症，却与身体功能失常没有任何明显的联系。让-马丁·沙可与约瑟夫·布洛伊尔两位医生都关注心理与行为之间的关系，而弗洛伊德在与他们合作的基础上，开始寻找其他可能性来解释他在一些病人身上所观察到的多种精神紊乱现象。

弗洛伊德开始相信，生理或身体机能方面的解释并不能理解和治疗这些精神疾病。他开始聚焦无意识心理，并使用梦的解析作为探索无意识的开创性方法。在弗洛伊德看来，"梦的构成涉及……种种精神病理学*问题。"[4]（"精神病理学"是指对心理—精神病症的研究）。弗洛伊德以他本人的梦和病人的梦作为基础提出了他的方法。他聚焦心理问题而不是生理问题。他通过解析梦境对心理问题进行评估的创造性方法，是弗洛伊德在《梦的解析》中所作出的独一无二的贡献。

时代贡献

在精神疾病学发展史上，弗洛伊德并非是第一个重视无意识心理因素的人。他与沙可、布洛伊尔的合作交流对他产生了深刻的影响。沙可打开了弗洛伊德的眼界，让他认识到他所观察到的一些行为问题可以追溯到患者早年的心理创伤。的确，"早期精神分析理论……显然要归功于弗洛伊德与沙可的交往。"[5] 弗洛伊德借鉴了沙可的思想，即童年发生的事件可能会对个体产生持久的影响——即使个体对此没有任何意识和感觉。

布洛伊尔通过与病人交谈进行治疗，这样的做法被证明是非常有效的。弗洛伊德对此亲眼目睹。他提出通过阐释和分析梦境来进行精神分析的观点时，这是一个重要的影响源头。布洛伊尔相信，恢复"对致病［歇斯底里症］事件的回忆……［可以］引导病人自由表达任何相关联的感觉以达到情感上的宣泄。"[6] "宣泄"是指净化或释放。

弗洛伊德通过增加释梦手段改进了这一治疗方法。他发现,病人在描述他们的梦境时,不太会抵制对无意识材料的讨论,尽管这些材料会带来某种不适。"在精神分析法的治疗过程中,我注意到人处于沉思时的心理状态与正在观察自我内心时的心理状态是截然不同的……人处于沉思状态时,具有批判力,其结果是拒绝接受他所感知到的一些思想……而另一方面,人处于自我观察状态时,其唯一的任务就是压制批判力。"[7]

---

1. 西格蒙德·弗洛伊德:《梦的解析》,A. A. 布里尔译,丹尼尔·T. 奥哈拉与吉娜·马苏奇·麦肯济导论和注释,纽约:巴恩斯和诺贝尔图书出版公司,2005年,第13页。
2. 弗洛伊德:《梦的解析》,第13—88页。
3. 弗洛伊德:《梦的解析》,第92—93页。
4. 弗洛伊德:《梦的解析》,第3页。
5. K. 利布雷希特与 J. 奎克宾:"论男性歇斯底里与心理创伤的早期历史:沙可对弗洛伊德思想的影响",《行为科学史学刊》第31卷,1995年第4期,第370—384页。
6. 理查德·韦伯斯特:"弗洛伊德、沙科与歇斯底里症:消失在迷宫中",登录日期2016年1月3日,http://www.richardwebster.net/freudandcharcot.html。
7. 弗洛伊德:《梦的解析》,第93页。

第二部分：学术思想

5 思想主脉

要点 🗝

- 《梦的解析》的中心主题是梦含有关于无意识心理中性欲望与侵犯欲的有意义的信息。
- 弗洛伊德花了大量时间讨论无意识心理如何通过凝缩*（将多种欲念压缩成一个元素）、置换*（用一物替换另一物）、表征化*（心理通过梦中内容呈现隐秘思想的方法）和润饰*（无意识心理为隐梦信息提供的叙事）来隐藏这些欲望。
- 弗洛伊德灵动的写作风格和对具体梦例的使用让他的观点更加有趣，更容易理解。

核心主题

西格蒙德·弗洛伊德在《梦的解析》中所提出的中心思想是：梦隐含着无意识心理欲望的重要信息。在弗洛伊德看来，这些欲望本质上是性的欲望。弗洛伊德引用社会学观点作为主要依据，说明无意识心理尤为关注"性的问题"，以及压抑性本能的问题。的确，他认为"没有任何其他本能从儿童时代起就受到如此之大的压抑；也没有其他本能遗留下如此之多、如此之强的无意识欲望；这些欲望在睡眠中处于活跃状态，随时都有可能形成梦境。"[1]

要承认无意识中性欲望的存在，有些人会感到不舒服。弗洛伊德相信，无意识心理中的很多部分都充当了审查官，遮蔽了性的内容，并以不太冒犯的方式在梦境中呈现出来。弗洛伊德区分了"显梦内容"（梦中呈现的"故事线"）和"隐梦内容"（梦的真实意义）。弗洛伊德认为，"就其重要性而言，隐梦内容要远远超过显梦内容。"² 换言之，只有揭开无意识中的隐秘欲望，我们才能发现梦的真实含义。要想如此，需要对梦进行分析和阐释。例如，一个抽雪茄的梦，实际上可能反映了做梦者痴迷男性生殖器的信息。

> "越是致力于寻找梦的答案，就越容易承认成年人的大多数梦都与性事有关，都是性欲的表现。"
> ——西格蒙德·弗洛伊德：《梦的解析》

思想探究

弗洛伊德相信，无意识心理主动地扭曲本能欲望。这种扭曲形成了梦的显性内容。具体地说，弗洛伊德认为这一主动扭曲的过程涉及四个方面：凝缩、置换、表征化与润饰。

"凝缩"是指意识可能将几种不同的欲念压缩成显梦内容中的一个元素。例如，弗洛伊德讲述了下面这个梦："我写了一本关于某种植物的专著。这本书摆在我面前。我随手翻动着一页折叠起来的彩色插图。每本书都有一片脱水的植物标本，好像来自植物标本册［收集植物标本的卷册］。"³ 在分析这个梦境时，弗洛伊德认为植物学著作同时指向几件事：他此前写过的关于古柯碱的著作，他的一位朋友曾经在实践中使用过古柯碱，名字叫弗洛拉的病人，他妻子最喜欢的花，他的研究，他的爱好——所有这一切都来自梦中的单一元素。弗洛伊德仔细审视了梦中的每一个元素，写出了令人晕眩的各种各样的指涉。他试图由此说明他的无意识选择了这些元素，是因为"它们能够表现出关于梦念的最广泛的联系，因此也代表了大量梦念聚集在一起的内核。"⁴

弗洛伊德还相信，无意识心理也在进行"置换"。换言之，梦中的内容可能是梦背后潜在欲念的替代物。弗洛伊德使用上述例子说明他的梦的意义之一是他在爱好上花钱过多的毛病。在这个例子中，"'植物学'这个元素无论如何不会在梦念的内核中占有一席之地，除非它与它的对应物存在某种松散的联系，因为植物学从来都不是我最喜欢的研究兴趣之一。"[5]弗洛伊德认为，他梦到自己不喜欢的爱好，说明他的无意识在指责他为了获得享受而在爱好上花了太多的钱。

弗洛伊德所说的"表征化"意思是指心理呈现隐梦内容的方式。在弗洛伊德看来，这通常会涉及到某种视觉意象。例如，一个人在塔顶上的意象可能代表"这个人的伟大"[6]，而大门的意象则"暗示身体上有一个口子"。[7]

无意识心理可以用来掩盖隐梦内容的最后一个方式是"润饰"。弗洛伊德认为意识对梦境进行了修饰，让其隐藏的各种信息难以辨识。换言之，梦境获得了连贯的故事线，故事线掩饰了深层的各种信息和矛盾。"梦的这部分内容是通过润饰得以完成的，对我们来说是清晰可见的；而润饰行为没能完成的那部分内容则是混沌不清的。"[8]

语言表述

为了向读者提供密切相关的背景信息，弗洛伊德举了很多范例来强化他的观点。读者也许不同意他对梦的本质所下的结论，但是读完这本著作后，会非常欣赏他的思考过程。

弗洛伊德是用德语进行创作的，因此在谈到他所使用的语言时，我们只能依据不同译者的翻译技能。奥地利精神病专家

A. A. 布里尔*，弗洛伊德的同代人，出版了《梦的解析》第一个英译本。不过，批评家们注意到布里尔的译著存在一些问题：他的译本中有些内容可能并不完全准确，他的英文表达流利，但是在某些地方存在舛误。[9] 大多数地方的语言直白晓畅，但是部分文本读上去太过拘泥形式，不像是现代文本；这反映了弗洛伊德时代的某些风气。例如，他把我们通常所说的"梦遗"*（引发男性遗精的梦）翻译成"梦污"。虽然语境有助于读者厘清大多数此类术语的含义，然而很多地方仍会让现代读者费解。此外，书中提到的其他科学家和哲学家的著作，布里尔都没有翻译出来。因此，读者必须掌握流利的西班牙语、法语，甚至拉丁语——或者借助于优秀的词典——才能有希望理解弗洛伊德书中的观点。

---

1. 西格蒙德·弗洛伊德：《梦的解析》，A. A. 布里尔译，丹尼尔·T. 奥哈拉与吉娜·马苏奇·麦肯济导论和注释，纽约：巴恩斯和诺贝尔图书出版公司，2005年，第245页。
2. 弗洛伊德：《梦的解析》，第146页。
3. 弗洛伊德：《梦的解析》，第268页。
4. 弗洛伊德：《梦的解析》，第269页。
5. 弗洛伊德：《梦的解析》，第287页。
6. 弗洛伊德：《梦的解析》，第320页。
7. 弗洛伊德：《梦的解析》，第322页。
8. 弗洛伊德：《梦的解析》，第400页。
9. 弗洛伊德：《梦的解析》，第IV页。

6 思想支脉

要点

- 《梦的解析》中的思想支脉包括儿童对父母产生的性冲动和书中第7章所论述的笼统的心理理论（此处指阐释心理机能本质的尝试）。
- 这些思想预演了弗洛伊德后期著作中俄狄浦斯情结论*（指儿童对相反性别的父亲或母亲的性欲望）和弗洛伊德关于性心理发展的"阶段论"（性快感的具体源头在个体的心理成长过程中扮演了重要的角色）。
- 弗洛伊德对心理理论的笼统描述为他的心理与人格结构的后期思想奠定了基础。

其他思想

西格蒙德·弗洛伊德在《梦的解析》中提出的核心思想得到了几个支脉思想的支撑。其中一个思想是弗洛伊德相信作为个人成长的正常组成部分，儿童对相反性别的父母会产生性依恋。这种依恋导致他/她们与同性别的父母形成竞争关系。弗洛伊德认为神经官能症的*患者（特定精神条件下出现焦虑症状的病人）所经历的许多症状都来自这一情结。他注意到，"父母在所有后来患上神经官能症的患者的婴幼儿*心理中起到了主要作用。"[1] 所以，当儿童表现出恋父憎母或恋母憎父的俄狄浦斯情结时，为他们将来的心理健康埋下了隐患。弗洛伊德认为他们产生了"心理冲动的消极因素，而这些心理冲动是在婴幼儿时期形成，对后来神经官能症*的发病起到了至关重要的推动作用。"²不过，他又补充道，"我并不认为神经官能症与正常人之间存在泾渭分明的差别。"³

弗洛伊德使用简单的术语来说明每个人都有这些感觉，神经官能症只是放大了这些感觉。再者，弗洛伊德认为对梦进行分析和阐释，可以深入了解这些问题。

这部著作的另一个思想支脉是弗洛伊德的心理机制论，特别是人的心理活动状态在清醒时和睡眠时的差异性。他把心理视作"一个复合型的工具，其各个构成部分可以称之为……机能。"⁴他用反射弧（当时通行的一个概念，指人的行为是对感官输入所作出的反射性回应）的语言来描述这些机能——感知。弗洛伊德说，"一般来说，心理过程的运作是从感知的一端到能动的一端"（"能动"是指运动）。⁵然而，就梦而言，"梦念的流动遭遇一系列的变形，我们不再把这些变形视之为正常的心理过程。"⁶

换言之，弗洛伊德相信，我们只有进行细致入微的分析和阐释，才能理解睡眠和梦中无意识心理的运作机制。

> "也许，我们所有的人都注定了把最初的性冲动指向我们的母亲，将最初的憎恨和暴力欲望指向了我们的父亲。我们的梦境证实了这一点。"
>
> ——西格蒙德·弗洛伊德：《梦的解析》

思想探究

在弗洛伊德看来，我们都会对父母产生冒犯性的性冲动。这个事实是无意识心理中有关性动力欲望的核心。为了展示这一过程的恒久性与普遍性,弗洛伊德回到古希腊,以其最著名的悲剧人物之一"俄狄浦斯"的名字来命名这个现象。在弗洛伊德重述的这个神话中,神谕告诉忒拜城的拉伊俄斯王*,他尚未诞生的儿子将来会弑父。当他的儿子俄狄浦斯出生后,拉伊俄斯王便要杀死这个孩子以阻止预言的实现——确切地说,他把孩子丢弃在野外,任其走向死亡。但是弃婴被人救走,最后流落到了国外,长大成人。后来,俄狄浦斯"遇见拉伊俄斯王,杀死了他……牵上了伊俄卡斯忒*的手"——他的母亲。7 因为这个神话起源于几千年前,弗洛伊德得出结论:俄狄浦斯情结经常发生。他以此作为证据,认为它是人类心理成长的正常组成部分。

在《梦的解析》第七章,弗洛伊德提出了他的心理学思想。他把人的心理描述为由多种机制或模块构成。有些模块直接接受感官信息,掌控运动神经行为。但弗洛伊德的理论还涉及心理机制中不太明显的部分。他说,梦本身"对了解心理机制的另一部分内容提供了证据。"8 他再次使用了读者所熟悉的反射语言("运动神经端"),并且这样写道:"在运动神经端的心理机制末端部分,我称之为前意识*……在这个机制的背后的那部分,我称之为无意识。"9 换言之,弗洛伊德相信,心理结构有一部分内容处于无意识与意识之间。这一极其重要的思想昭示着他在后期作品中将人的心理结构分为三个组成部分,即著名的本我*、自我*与超我*。本我处于无意识的中心,是原始本能,如性欲;自我是心理结构中有意识的部分,我们形成社会身份的那部分;超我是一个人以良知*意识和道德感等为中心的那部分。

在后来的著作中,弗洛伊德使用了《梦的解析》中的一些基本思路,论述了性心理成长中的"阶段论"思想——即从婴幼儿到成年人的成长过程中，我们经历各个不同的阶段，并从身体的不同部位获得性快感，而这一过程会给未来人生带来不同的影响。

被忽视之处

《梦的解析》给精神病学家和心理学家*思考心理结构与治疗精神疾病的方法带来了意义重大的变化。作为一部影响深远的著作，它多年来受到了大量批评者和支持者的关注。因此，很难想象该书提出的思想也受到了不应有的忽视。

不过，我们可以认为弗洛伊德的著作在某种程度上被现代世界所忽视了。的确，近年来，人们发现"心理学教科书中精神分析学被边缘化了"。[10]弗洛伊德的思想被摒弃，因为弗洛伊德没有提供实验性的*证据来证明他的论断（也就是说，没有经过观察和验证过的证据）。这一边缘化最近有加速的倾向，因为一些研究心理健康的专业人士已经将重心转移到精神疾病的生理层面上。我们无法否认的是，总体来看，弗洛伊德在心理学领域以及在人类世界留下了不可磨灭的印记。不过，我们也可以说，他在《梦的解析》中所提出的大多数具体观点并未对当代心理学家和精神病专家产生很大的影响。[11]不过，我们发现，认知心理学家*和神经科学家们*（研究大脑和神经系统的科学家）开始对弗洛伊德的某些思想产生了越来越多的兴趣。他们对大脑在意识感觉之外如何储存和使用特定类型的记忆兴趣浓厚。

1. 西格蒙德·弗洛伊德:《梦的解析》,A. A. 布里尔译,丹尼尔·T. 奥哈拉与吉娜·马苏奇·麦肯济导论和注释,纽约:巴恩斯和诺贝尔图书出版公司,2005年,第226页。
2. 弗洛伊德:《梦的解析》,第226页。
3. 弗洛伊德:《梦的解析》,第226页。
4. 弗洛伊德:《梦的解析》,第424页。
5. 弗洛伊德:《梦的解析》,第424页。
6. 弗洛伊德:《梦的解析》,第467页。
7. 弗洛伊德:《梦的解析》,第226—227页。
8. 弗洛伊德:《梦的解析》,第427页。
9. 弗洛伊德:《梦的解析》,第428页。
10. 约瑟夫·里彭:"西格蒙德·弗洛伊德对于21世纪的意义",《精神分析心理学》第23卷,2006年第2期,第215—216页。
11. 保尔·R. 麦克休:"弗洛伊德之死与精神病学之重生",《标准周刊》,登录日期2016年1月10日,http://www.weeklystandard.com/article/12226。

7 历史成就

要点 🗝

- 弗洛伊德聚焦神经症患者和健康个体的无意识心理的作用，对精神病学和心理学产生了深刻的影响。
- 弗洛伊德的创新思想和轻松愉快的写作风格吸引了大量的临床医生，其结果是创立了精神分析学这一引人瞩目的思想流派。
- 弗洛伊德的观察所得带有八卦性（这些观察的"故事性"只适用于某些个体），其理论缺少实验性——或者实证性的支持，因此，他在《梦的解析》中所提出的观点的影响力受到了限制。

观点评价

在《梦的解析》中，西格蒙德·弗洛伊德旨在表明梦作为无意识心理的产物，"仅仅是欲望的实现，而没有其他任何目的。"[1] 他描述了梦的解析方法，指出释梦是洞悉无意识心理的手段。弗洛伊德认为，洞悉无意识中的欲念，可以为临床专业人士治疗不同的心理疾病提供另一种可行性的途径。这样做还可以提升做梦者的自我感知能力。

弗洛伊德将他的方法部分建立在他和奥地利医生约瑟夫·布洛伊尔合著的《癔症研究》[2]（1895）中所描述的治疗方法上。在该书的一节中，一位病人对梦作出描述，并采用自由联想法来解释梦的真实含义。弗洛伊德相信，这个治疗过程帮助病人改善了症状。他自述采用这个方法获得了巨大的成功。他由此指出，"只要从病人精神生活的各种因素中成功追溯到某个病态观念的发病源头，那么这个观念就会土崩瓦解，病人也就能获得痊愈。"³ 然而，弗洛伊德这些推论的不科学性在后来很多年里引起了人们的批评。

尽管受到批评，弗洛伊德在《梦的解析》中所提出的观点最终在精神病学和心理学领域创立了引人瞩目的精神分析学派。弗洛伊德由此实现了他的主要目标：他让人注意到了他的观点，即梦向我们传达了无意识心理中的信息。

> "没有其他的证据表明，弗洛伊德对精神病学的巨大影响主要在于研究视野的转变，这些影响主要归功于他的著述。"
>
> ——E. 斯坦格尔："弗洛伊德对精神病学的影响"

当时的成就

弗洛伊德的心理理论比他的先驱们的思想要"更加深刻，更加准确"⁴，由此吸引了来自临床医学界的众多专业人士。这本著作的第二版几乎是在10年后才得以再版。不过，此后几十年中，《梦的解析》广受青睐，印数激增。事实上，仅在1910至1929年间，出版商就加印了6次之多。⁵

弗洛伊德因为在精神病学和心理学领域创立了精神分析学派而被许多人所称颂。精神分析学着重强调无意识心理在影响意识行为和人格结构中的作用。它在20世纪早期产生了巨大的影响。我们甚至可以说，20世纪中叶之前，精神分析学始终是该领域最引人瞩目的理论视角之一。⁶一位作家曾经这样回忆道：在他的研究生学习期间，"弗洛伊德是研究热点，心理学系的教师们许多都是精神分析学家。"⁷

弗洛伊德用如此有趣的方式所提出的创造性思想，吸引了大量读者，很多读者后来成了他的信徒。他的关于无意识心理与通过释梦方式洞悉无意识心理的思想，与当时的主流思想形成了鲜明的对比。瑞士思想家卡尔·荣格曾经是弗洛伊德的学生，后来凭一己之力成为很有影响的精神病专家。他说，"精神分析理论的创立，是科学史上一件非同寻常的大事，具有巨大的优势，因为它以其大胆的思想而成为特立独行、独树一帜的现象，而且在哲学与科学背景的映衬下，显得尤为引人夺目。"[8]弗洛伊德本人也承认自己在改变临床医学领域看待无意识的方式上所取得的成就。在这本著作第三版的前言中，他写道："释梦必定会有助于展开对神经官能症的心理分析。"[9]那些提供弗洛伊德式新交谈疗法的治疗师们曾经显赫一时，尤其是在美国。

局限性

尽管《梦的解析》大获成功，但是其局限性最终限制了它的传播范围，削弱了它的影响力。这本著作起初销量不佳，让弗洛伊德感到不安，这是人之常情。在这本书的第二版前言中，他写道："如果这本书的第二版还能重印……我对那些我原本企望会感兴趣的临床专业人士们不存感激。"[10]然而，这本著作后来不断再版，说明读者兴趣激增。

《梦的解析》最根本的缺陷在于弗洛伊德提出无意识学说的方式。他依赖自我分析与逸闻趣事来构想他的理论，而不是借助系统的、实验性的（以实证为基础的）观察所得。17和18世纪的文化运动，即著名的启蒙运动强调理性、逻辑与科学调查的重要性。弗洛伊德的同代人赞同这些价值原则，而弗洛伊德带有八卦性、不作验证的理论似乎抛弃了这些价值原则。他的著作的正确性因此受到了质疑。的确，有人将弗洛伊德看成是"用心不良的恶棍，或者是一个长着病态大脑的家伙，用自己的妄想来冒充临床医学的观察结果。"[11]

---

1. 西格蒙德·弗洛伊德：《梦的解析》，A. A. 布里尔译，丹尼尔·T. 奥哈拉与吉娜·马苏奇·麦肯济导论和注释，纽约：巴恩斯和诺贝尔图书出版公司，2005 年，第 446 页。
2. 西格蒙德·弗洛伊德和约瑟夫·布洛伊尔：《癔症研究》，詹姆斯·斯特拉齐译，伦敦：霍加斯出版社，1955 年。
3. 弗洛伊德：《梦的解析》，第 92 页。
4. 厄尼斯特·琼斯："弗洛伊德与他的成就"，《英国医学杂志》第 1 卷，1956 年第 4974 期，第 997—1000 页。
5. "弗洛伊德的新作《梦的解析》1900 年问世"，PBS（美国公共广播公司），登录时间 2016 年 1 月 24 日，http://www.pbs.org/wgbh/aso/databank/entries/dh00fr.html。
6. "心理学引论"，《心理学》，1—34，德克萨斯州休斯顿：免费教科书计划，2014 年。
7. 约瑟夫·里彭："西格蒙德·弗洛伊德对于 21 世纪的意义"，《精神分析心理学》第 23 卷，2006 年第 2 期，第 215—216 页。
8. C. G. 荣格："历史语境中的西格蒙德·弗洛伊德"，《人格学刊》第 1 卷，1932 年第 1 期，第 48—55 页。
9. 弗洛伊德：《梦的解析》，第 7 页。
10. 弗洛伊德：《梦的解析》，第 5 页。
11. 琼斯："弗洛伊德与他的成就"，第 998 页。

8 著作地位

> **要点**
>
> - 弗洛伊德一生致力于探索无意识心理在人类行为与人格结构中的重要意义。他由此在心理学和精神病学领域创立了精神分析学派。
> - 《梦的解析》为弗洛伊德的后期理论奠定了基石。
> - 很多人视《梦的解析》是弗洛伊德最重要的学术成就之一。该书为他赢得了精神分析学之父的美誉。

定位

在撰写《梦的解析》之前,弗洛伊德最出名的著作是1895年出版的《癔症研究》,由他与奥地利医生约瑟夫·布洛伊尔合作完成。[1] 批评家们视之为"精神病理学的里程碑式著作",认为它的出版日期是"精神分析学的肇始"。[2] 在这本书中,弗洛伊德和布洛伊尔描述了临床个案研究,记载了独特的治疗方法。他们新尝试的"交谈疗法"[让病人讨论自己的问题],被证明可以缓解精神病的病症。这一疗法的成效极大地激发了弗洛伊德的兴趣,使他感到有必要"沿着布洛伊尔开辟的道路勠力向前,直到这个课题得到全面的理解。"[3]

5年后,弗洛伊德出版《梦的解析》。这部著作反映了他在与布洛伊尔合作期间与合作之后所形成的思想。对于无意识心理在人类行为与人格发展过程中的作用,弗洛伊德始终兴趣不减。这本著作用释梦方式来探究无意识心理中的欲望,主要得益于他的这一兴趣。《梦的解析》为他后期许多专业性著作奠定了基础。书中提出的几个观点预演了弗洛伊德后期著作中的两大理论：心理理论（关于心理的本质、结构与功能的理论）和性心理发展的"阶段论"——根据"阶段论"，我们的心理成长从婴幼儿到成年人要经历几个不同的阶段，在不同的阶段我们会从不同的身体部位获得性快感。

> "弗洛伊德在构想出一套方法，即现在人所共知的精神分析学方式后，他一生的工作在广义的层面上可以被归结为是对无意识的探索。"
>
> ——厄尼斯特·琼斯："弗洛伊德与他的贡献"

整合

纵览其作为精神分析学家的一生，弗洛伊德相信，为了理解人类行为与人格结构，我们必须理解无意识心理中的欲望。弗洛伊德的所有著作都以这个观点作为中心命题，以《梦的解析》为奠基石。在《梦的解析》中，弗洛伊德对人的心理作出了笼统的、反应式的描述。在后期作品中，他概述了"心理的'地形学'，将它分为三个部分：本我、自我和超我。"[4]他将这些内容界定为无意识心理的不同分区：它们同时隐含着个体心理所压抑的所有欲望和愿望。弗洛伊德相信大多数欲望在本质上属于性欲。自我基本代表了意识心理部分。它被"夹在本我与超我之间。"[5]超我代表了良知意识。本我代表无意识心理的性欲和侵犯欲。根据弗洛伊德的理论，自我始终处于紧张状态，试图平衡本我与超我的不同需求。

在《梦的解析》中，弗洛伊德暗示了另一个重要理论的到来。在描述儿童对父母产生的冒犯性的性冲动时，弗洛伊德引用了古希腊神话俄狄浦斯王*的故事。弗洛伊德在提出性心理发展的阶段理论时，这个故事在他的后期著作中占有一席之地。这些阶段包括弗洛伊德取名为"俄狄浦斯情结"的现象。他将这个现象看作是每个人心理发展过程中的正常部分。简言之，性心理发展的阶段理论是指性心理的发展跨越了儿童不断成长的身体中不同的性感区*（可以获得性快感的身体部位）。

弗洛伊德将每一个快感区与一系列的冲突联系在一起；而人如果不能成功消除这些冲突，就会形成影响人格与行为的固恋*（极度的迷恋）。例如，最早的阶段之一是以口腔为中心的快感区。这个阶段与婴儿的早年生活相关，婴儿一般靠吮吸乳汁而存活。如果母亲断奶太早或太迟，婴儿可能会出现口腔固恋症。同样，"成年人抽烟、喝酒、暴食或咬指甲"[6]，都会被认定是得了口腔固恋症。随着时间的推移，快感区从口腔转移到肛门，然后又转移到生殖器或阴茎（俄狄浦斯情结发生期）。在性器期之后，则是一段相对的静止期。此后，大约在青春期，快感区域主要以生殖器为中心，并一直持续到个体生命的未来。

意 义

弗洛伊德是一位多产的作家，一生出版了300多部作品。不过，他经常将《梦的解析》视作个人最喜爱的著作。[7]的确，"《梦的解析》第3个英译本（1931）的前言清楚表明这本书在作者眼中的地位：'在我有幸所作出的所有发现当中，［这本书］含有……最有价值的发现。一个人有幸获得如此重要的洞察力，一生当中仅只有一次。"[8]他的支持者与批评者都认同这个观点，习惯性地称"释梦一书是精神分析学的核心著述"[9]。支持者们搜罗"赞美性修饰语［称赞的话语］，如［德裔美国历史学家］彼得·盖伊*将它与［进化论奠基人］查尔斯·达尔文的《物种起源》相媲美，称它是'形塑现代文化的具有革命性的经典之作'"。[10]

在职业生涯早期，弗洛伊德出版了很多著作，所研究的论题从古柯碱的历史与临床应用到儿童脑瘫（脑性瘫痪）。[11] 在《梦的解析》之后，他的著作沿着同一个方向发展，继续深入论述他在这部杰作中提出的很多观点。《梦的解析》是弗洛伊德的一部"巨著［杰作］……它是弗洛伊德后期著作的基石。"[12] 在最近几十年，尽管弗洛伊德的影响逐渐式微，但是这部著作在他的职业生涯中的中心地位并没有减弱。

---

1. 西格蒙德·弗洛伊德和约瑟夫·布洛伊尔：《癔症研究》，詹姆斯·斯特拉齐译，伦敦：霍加斯出版社，1955 年。
2. 厄尼斯特·琼斯："弗洛伊德与他的成就"，《英国医学期刊》第 1 卷，1956 年 4974 期，第 997—1000 页。
3. 西格蒙德·弗洛伊德：《梦的解析》，A. A. 布里尔译，丹尼尔·T. 奥哈拉与吉娜·马苏奇·麦肯济导论和注释，纽约：巴恩斯和诺贝尔图书出版公司，2005 年，第 92 页。
4. 保尔·利科："西格蒙德·弗洛伊德"，载《保尔·利科：劳特利奇批评思想家》，卡尔·西蒙斯，英国阿宾登：泰勒-弗朗西斯出版集团，2002 年，第 46 页。
5. 利科："西格蒙德·弗洛伊德"，第 46 页。
6. "人格"，载《心理学》，369—410，德克萨斯州休斯顿：免费教科书计划，第 375 页。
7. 肯德拉·车里："弗洛伊德的著作：弗洛伊德最著名与最有影响力之作"，登录日期 2016 年 1 月 9 日，http://psychology.about.com/od/sigmundfreud/tp/books-by-

sigmund-freud.html。
8. 帕特里夏·凯切尔:《弗洛伊德的梦：跨学科的心理科学全景》，麻省剑桥：麻省理工学院出版社，1992 年，第 113 页。
9. 凯切尔:《弗洛伊德的梦》，第 113 页。
10. 凯切尔:《弗洛伊德的梦》，第 113 页。
11. 琼斯:"弗洛伊德与他的成就"，第 997—1000 页。
12. 琼斯:"弗洛伊德与他的成就"，第 999 页。

第三部分：学术影响

9 最初反响

> 要点 🗝
>
> - 弗洛伊德对婴幼儿性欲的重视,以及他的非科学的研究方法,颇受早期批评者们的诟病。
> - 弗洛伊德忽视了部分批评。但当曾经的同事开始批评他的著作时,他往往中断与他们的联系。
> - 读者如何理解他的观点,往往取决于读者的个人信念以及他们对弗洛伊德的忠诚度。

批评

西格蒙德·弗洛伊德的《梦的解析》出版后的最初几年,批评家们大多未加理睬,因为它并不符合当时普遍的科学主题。然而,它的受欢迎程度不断增加。弗洛伊德的思想强势猛进,从而在精神病学与心理学领域引领了一个新的思想潮流:精神分析运动。第一届精神分析学国际代表大会于1908年召开。两年后,一批精神分析学家创立了国际精神分析学会。弗洛伊德成为学界名家。他和他的学生、瑞士精神分析学家卡尔·荣格还奔赴美国讲授精神分析学。[1]

尽管这本著作赢得了这些赞誉,但也引起了不少争议和批评。弗洛伊德的批评者们对他仅凭主观推测而提出相关观点表示反对。他们把这一点当做是主要缺陷,由此对弗洛伊德著作的科学价值提出质疑。有些人只是与弗洛伊德疏远了关系,而另外一些人则公开批评他的观点。

曾与弗洛伊德合著《癔症研究》(1895)的约瑟夫·布洛伊尔既与他疏远了关系，也批评了他。的确，学界一般认为，布洛伊尔切断了与弗洛伊德的一切联系，"因为布洛伊尔反对弗洛伊德将性看成是精神神经病的病因*。"[2]（"病因学"是指对疾病起因的研究。）还有人声称，他们俩断绝了关系，是因为"布洛伊尔……无法忍受［弗洛伊德思想中固有的］对理性主义的巨大侮辱。"[3]

从曾经的同事转变成弗洛伊德的批评者，布洛伊尔并不是唯一的人。卡尔·荣格，弗洛伊德的学生，后来也成为弗洛伊德尖锐的批评者。"他们俩的关系经历了前亲后疏的转折：弗洛伊德起初宣称荣格是他的'继承者'，6年后认为他'疯了'。"[4] 弗洛伊德认为，性本能与儿童早年经验在成年后的性格中起到了重要的作用。当荣格拒绝接受这一观点时，两人关系开始决裂。荣格指出，人的各种目的与抱负也是人类行为的重要推动力。荣格认为，弗洛伊德关于无意识心理的思想还不够成熟。荣格的个体无意识思想虽然与弗洛伊德的看法十分相似，但荣格发现了人类无意识的另一层，他称之为"集体无意识"*[5]（每个个体都拥有和分享的无意识部分）。荣格与他从前的老师存在的另一个差异是他对待宗教与魂灵的态度。他将宗教与魂灵归入集体无意识。[6]

> "弗洛伊德的理论……至多带有部分真理性，因此为了维护自身及其有效性，它犹如教条一样死板，宗教审判人一样狂热。"
> ——卡尔·荣格："历史语境中的西格蒙德·弗洛伊德"

回应

一位作家在描述弗洛伊德对这些批评的回应时指出，他"用巨大的坚韧承受住了所有的敌意，从没有俯身屈就对敌意作出公开的回应；私下里，他甚至还以此为乐……我记得他曾经这样说过：'我的对手们可能在白天攻击我的学说，但是我相信，他们夜间会梦见我的学说。'"[7] 他置身事外的态度并没有弱化对手们的批评。当时，批评者们不断对弗洛伊德理论的不科学性提出了很多相同的质疑。

弗洛伊德可以用幽默化解某些对手们的批评，但是那些来自同事的批评，那些曾经是他最亲密的盟友的批评，深深伤害了他。荣格的批评让弗洛伊德感到自己被背叛了，内心很受伤害。两人早年是真挚好友，相互尊重，最后变成了互不相让的竞争对手。荣格与弗洛伊德的通信往来变得越来越敌对。最后，他们同时中断了联系。在写给荣格的最后一封信中，弗洛伊德说，"我提议，我们彻底中止我们的个人关系。这对我来说不会有任何损失，因为我与你的情感联系早就变成了一根细线。"[8] 此后，弗洛伊德与他的批评者们割断了联系。而他的周围聚集了一批衷心维护精神分析学说的同事，他们坚决维护精神分析学派的原则。[9]

冲突与共识

弗洛伊德与他的批评者们分歧很大，这些分歧持续了很多年。能够感受到弗洛伊德专业与个人怒火的，不仅仅是荣格与布洛伊尔这两位前同事。弗洛伊德还与奥地利医生、精神治疗师阿尔弗莱德·阿德勒*、匈牙利精神分析学家桑多尔·费伦齐*决裂。这两人曾经都是他的学生。精神分析学家厄尼斯特·琼斯*自始至终都是弗洛伊德的支持者。他说，阿德勒和费伦齐无法"接受深层次的精神分析调研后所得出的结论……他们掉头而去，摒弃了他们曾经信服与阐释过的观点。"[10] 然而，琼斯承认，同阿德勒与荣格一样，许多批评家"在拒绝弗洛伊德的性理论后，能够享受职业上的成功，但是普通大众对他们从弗洛伊德令人不悦的思想中获救而如释负重，并为之欢呼。"[11]

总而言之，对弗洛伊德《梦的解析》的反应主要取决于个人信念与忠诚度。有些人，例如琼斯，自始至终、坚定不移地忠实于弗洛伊德。后来，琼斯成了弗洛伊德传记的撰写者。一些理论家像荣格一样，接受了弗洛伊德的很多思想，但是对这些思想进行了改进，以适用于他们自己对无意识心理的本质所提出的观点。布洛伊尔和其他人彻底抛弃了弗洛伊德的思想。他们反对弗洛伊德建构无意识理论与梦的解析的不科学方法。

---

1. 西格蒙德·弗洛伊德：《梦的解析》，A. A. 布里尔译，丹尼尔·T. 奥哈拉与吉娜·马苏奇·麦肯济导论和注释，纽约：巴恩斯和诺贝尔图书出版公司，2005 年，第 xv 页。
2. 约翰·P. 穆勒："重读《癔症研究》：重访弗洛伊德—布洛伊尔的决裂"，《精神分析心理学》第 9 卷，1992 年第 2 期，第 129—156 页。
3. 穆勒："重读《癔症研究》"，第 129 页。
4. 海斯特·麦克法兰·所罗门："弗洛伊德与荣格：一次不完整的相遇"，《分析心理学学刊》第 48 卷，2003 年第 5 期，第 553—569 页。

5. 肯德拉·车里:"西格蒙德·弗洛伊德自画像:弗洛伊德与荣格",登录日期 2016 年 1 月 9 日, http://psychology.about.com/od/sigmundfreud/ig/Sigmund-Freud-Photobiography/Freud-and-Jung.html。
6. "卡尔·荣格与西格蒙德·弗洛伊德证据充分的友谊",历史甲骨文网站,登录日期 2016 年 1 月 9 日, http://historacle.org/freud_jung.html。
7. 厄尼斯特·琼斯:《弗洛伊德与他的成就》,《英国医学杂志》第 1 卷,1956 年第 4974 期,第 997—1000 页。
8. 戴维·艾登伯格:"弗洛伊德和荣格:往来信函中的'精神分析学'",《心理学视角》第 57 卷,2014 年,第 7—24 页。
9. 车里:"西格蒙德·弗洛伊德自画像:弗洛伊德与荣格"。
10. 琼斯:《弗洛伊德与他的成就》,第 998 页。
11. 琼斯:《弗洛伊德与他的成就》,第 998 页。

10 后续争议

要点 🗝

- 《梦的解析》使精神分析学成为一种治疗方法与一大思想流派。
- 精神动力学方法（将无意识心理与有意识行为联系起来的方法）如同分析心理学视角*一样，是在弗洛伊德著作的启发下创立的。
- 学界将《梦的解析》视作弗洛伊德最伟大的著作；任何人想要更好理解精神分析学的基本原理，这部著作仍然是核心读本。

应用与问题

许多批评者对西格蒙德·弗洛伊德的《梦的解析》不屑一顾。他们认为，弗洛伊德"夸大了性的作用，依托错误的方法，屈服于疯狂的推测。"[1] 然而，他的支持者们认为——始终认为——他的著作对精神病学与心理学产生了巨大的影响。

《梦的解析》奠定了精神分析学的基石，确立了弗洛伊德在精神分析运动中毋庸置疑的领导地位。弗洛伊德的视角创立了一种崭新的思维方式——"引发了一场文化革命，在广度上可以与［英国自然学家和进化论奠基人］查尔斯·达尔文相媲美。"[2] 它还为专业医疗人士提供了崭新的治疗方法。本书出版后不久，很多精神病专家开始采用梦的解析与自由联想等方法来治疗不同类型的精神病人。[3]

弗洛伊德始终强力维护自己理论中的主要观点。他坚持认为，性本能对无意识心理中的各种欲望起到了激发的作用。对他来说，这是一个核心的观点。弗洛伊德对自己观点的固执坚守，导致他与最坚定的支持者关系紧张。很多支持者们很想改进他的无意识理念。弗洛伊德的学生卡尔·荣格以及其他人,最终创立了他们自己的相关思想流派。他们从弗洛伊德的思想出发,后来被称作"新弗洛伊德派",并凭着各自的努力成为卓越的专业人士。

> "弗洛伊德不仅拓宽了精神病学的疆域,而且还赋予其新的维度,主要借助于对动力无意识——即影响人类行为和意识的精神力量——的深刻发现与探索。"
> ——E. 斯坦格尔:"弗洛伊德对精神病学的影响"

思想流派

精神分析学诞生于弗洛伊德的著作,强调无意识心理对行为与人格的影响。弗洛伊德将无意识看作是未实现欲望与本能的水库。人往往在梦中释放他们的无意识性欲望。实际上,在他看来,与无意识相关的大多数心理能量直接来自性本能,而性本能被某种审查机制从意识感觉中驱赶出去。弗洛伊德后来将这种内在审查称作"超我"(掌控道德与我们无意识本能的那部分心理)。弗洛伊德认为,我们只有通过对梦进行分析与阐释,才能真正理解这些无意识欲望。[4]

如同很多创新者所遇到的那样,不是所有弗洛伊德的追随者们都全盘接受他的思想。一些理论上的分歧促成了不同思想流派的诞生。学界赞赏弗洛伊德的学生兼批评者卡尔·荣格创立的分析心理学说——精神分析学的另一支思想流派。荣格提出他自己独树一帜的集体无意识思想(我们与他人共同拥有的无意识心理)以及共同拥有的意象和模块,他称之为"原型"*。与弗洛伊德不同的是，荣格还将他的许多思想运用到宗教与魂灵层面。[5]如果弗洛伊德作为老师能改进他的方法，荣格可能会一直是他的支持者。然而，"弗洛伊德将荣格钦定为接班人，期待他能全盘接受自己所构想的精神分析理论，但他对这个年轻的精神病医生的判断严重失误。"[6]

当代研究

当代学者经常将精神分析学、分析心理学以及其他学说划为一派，冠名为"精神动力学派"。精神动力学理论是指"那些基于内心各种驱动力和力量，特别是无意识力量相互作用，以及不同人格结构相互作用的理论来探讨人体机能的理论。"[7]近年来，特别是精神分析学的影响日渐式微，许多精神病专业人士仍然运用弗洛伊德和他的追随者们（即新弗洛伊德派）提出的治疗方法。如果有兴趣了解这一学说，阅读这本著作不失为明智之举。有一位临床医生写道，"如果对'弗氏精神分析学'的基本思想与理论发展不作深入了解，不作深刻的研究，我就不可能成为全面的'荣格式精神病医师'。"[8]

在临床医疗范围之外，人们依然广泛讨论《梦的解析》与弗洛伊德的其他著作。不过这些讨论一般集中在历史价值与影响上。认知神经科学*（对思维过程的科学研究）所取得的最新进步，为弗洛伊德的著作在当今更加突出的作用打开了一扇大门。一位作家注意到，"在特定的精神病学与神经科学界的语境下，人们对精神分析学的兴趣历久弥新。"[9]的确，早在20世纪80年代中叶，学者们就开始为"精神分析学理论和实践的潜在合法性"进行辩护。[10]

1. 安东尼·D. 康得斯:"真理、真理性与精神分析学:弗洛伊德在德国威廉二世时代的接受",《德国历史》第 31 卷,2013 年第 1 期,第 1—22 页。
2. 亨利·F. 艾伦伯格:《无意识的发现:动力精神病学的历史与演化》,纽约:基础图书公司,2008 年,第 418 页。
3. 厄尼斯特·琼斯:《弗洛伊德与他的成就》,《英国医学杂志》第 1 卷,1956 年第 4974 期,第 997—1000 页。
4. 索尔·麦克劳德:"精神动力学方法",《简易心理学》,登录日期 2016 年 1 月 10 日,http://www.simplypsychology.org/psychodynamic.html。
5. 麦克劳德:"精神动力学方法"。
6. 海斯特·麦克法兰·所罗门:"弗洛伊德与荣格:一次不完整的相遇",《分析心理学学刊》第 48 卷,2003 年第 5 期,第 554 页。
7. 麦克劳德:"精神动力学方法"。
8. 所罗门:"弗洛伊德与荣格",第 553—569 页。
9. 尼玛·巴西里:"弗洛伊德与脑物质:论神经精神分析学的重组",《批判性探索》第 40 卷,2013 年第 1 期,第 83—108 页。
10. 巴西里:"弗洛伊德与脑物质",第 83 页。

11 当代印迹

要点 🗝

- 一方面,《梦的解析》对当下的精神健康专业人士来说越来越无关紧要,而另一方面,它仍然是一个经典文本,能激发现代心理学家们的兴趣。
- 弗洛伊德对无意识重要性的强调,让人们关注到人类心理的无意识层面,已经在心理学的历史中留下了不可磨灭的印记。
- 关于弗洛伊德思想与 21 世纪的相关性,学界争议不断。有些人将他的观点降格到了故纸堆的层面,有些人则认为这些思想在当今仍然具有潜在的价值。

地位

西格蒙德·弗洛伊德的《梦的解析》开启了精神分析运动。"从 1935 年到 1975 年……弗洛伊德主义是美国精神病学不可挑战的信条。"[1] 一位作家哀叹道:"近年来生物医学界*有很多发现,普通大众可能视之为巨大进步,但这些发现其实掏走了精神病学的'灵魂',使它成了一桩冰冷的生意,只能向病人们配发药片,而不能针对患者可悲的独特病症进行疏通开导。"[2]("生物医药法",治疗精神疾病的方法,主要立足于治疗患者的生理机能,通常依赖药物治疗。)有人提出这一变化让精神病学"成长为一个以科学为基础、以证据为驱动的学科。"[3]

今天,治疗精神病的专业人士采用广泛的治疗方法。精神分析治疗师们仍然提供治疗服务,但是他们的人数,还有他们的影响,已经呈稳定的下降趋势。据称,"出版精神分析学著作的出版商急剧减少……主流的大学出版社甚至不再愿意出版精神分析学著作。"[4]

有人将这一态度上的变化归咎于"精神病学中的生物学转向与认知神经科学的崛起,后者成为世界各国心理学系和学术期刊的主导范式。"[5] 认知神经科学根据人脑的电子和化学机能来探讨人的思维。尽管它似乎敲响过精神分析学的丧钟,但它也有可能是精神分析学复活的良机。随着人们对神经科学的兴趣"在过去几十年中稳步增长……一些从业者们越来越关注的探索区域,属于或接近精神分析学的研究领地。"[6] 精神分析学与神经科学可以成为潜在的合作学科,在共同探索诸如意识感觉之外的记忆方面存在可能性的机遇。

> "有人认为,在学科交叉融合的过程中,精神分析学与神经科学能够直接惠及对方;神经科学可以与某种更具活力的主观经验理论相融合,而精神分析学可以过渡成为一门可验证的实验科学。"
> ——尼玛·巴西里:"弗洛伊德与脑物质:论神经精神分析学的重组"

互动

不是所有人都热衷于将精神分析学和神经科学交叉融合的想法。其实,精神分析学阵营中的很多人持抵制态度,将精神分析学"视作知识的'非科学领域',与人文学科*密切相关,远远超出了自然科学"。[7] "人文学科"是一个范围很广的研究领域,包括历史学、文学和文化,因此,有人觉得将精神分析与神经科学联系在一起,它的存在基石会受到威胁。

尽管如此,精神分析学这一领域仍然四分五裂。的确,有人认为"大量的团体和派系……对弗洛伊德学说的真正遗产莫衷一是;颇为不幸的是,他们没有投入足够的时间和努力来提升教学与研究方法。"[8]因此,"在很多国家,精神分析学已经从学术前沿消失,特别是在美国,精神分析学曾经在许多心理学系占据优势地位。"[9]

许多精神分析学者甚至开始质疑弗洛伊德思想与21世纪的相关性。一位作家写道,"存在'两个弗洛伊德'……自然科学家*弗洛伊德和阐释学家弗洛伊德。"[10]阐释学家是指擅长于文本解读的专家——换言之,我们既可以将弗洛伊德看成是一位临床医生,也可以将他看成是阐释学的专家,具体说来,阐释个体通过梦境所表达的未实现的欲望。这位作家认为,要想确定弗洛伊德思想的相关性,我们必须在个案的基础上加以探讨。

持续争议

神经科学可能探讨储存在无意识心理中的记忆。许多精神分析学家对这种可能性持欢迎态度。将精神分析学原理运用在神经科学的语境中,可以让我们"研究精神生活的生理支撑,有助于补充和完善我们的思想,另一方面也为我们提供各种模型和概念,用以丰富生物学家们的研究。"[11]换句话说,今天的技术手段可以向我们展示无意识过程如何产生了人脑中的活动。

然而,神经科学家们探讨这些思想所采用的方法,与他们的精神分析学同行们截然不同。正如一位学者断言,"弗洛伊德以生物学为基础所提出的思想……与21世纪没有关系……事实证明,精神分析学家已经开始关注生理机能对心理反应的影响,但是,对弗洛伊德的思考起指导作用的,是建立在神经科学与人脑结构基础上的生理学,而不是以研究神经机能与心理的遗传内容为主的神经生理学*。"[12]"神经生理学"此处是指对人脑与神经系统中生理结构与机能的研究。

尽管如此，神经科学可以为精神分析学提供始料未及的生机。如果这个学科不确立它与当下的相关性，它将"不可避免地被淘汰，这是它无法或拒绝发展成自然的实验性科学的结果。"[13] 我们可能发现此处所存在的反讽：弗洛伊德是一个科班出身的医学博士，但是在创立精神分析学说时，却断然抛弃了以检测和验证为基础的科学方法。相反，他所依赖的是八卦和个人经验。如果精神分析学想要保持与当下的相关性，它从现在开始就必须要依赖科学。

---

1. 保尔·R.麦克休："弗洛伊德之死与精神病学之重生"，《标准周刊》，登录日期2016年1月10日，http://www.weeklystandard.com/article/12226。
2. 麦克休："弗洛伊德之死与精神病学之重生"。
3. 麦克休："弗洛伊德之死与精神病学之重生"。
4. 卡罗·斯特伦格："精神分析学能够重回公共领域吗？"，《精神分析心理学》第32卷，2015年第2期，第293—306页。
5. 斯特伦格："精神分析学能够重回公共领域吗？"，第294页。
6. 米格尔·安琪儿·冈萨雷斯-托雷斯："精神分析与神经科学：朋友，还是敌人？"，《国际精神分析学论坛》第22卷，2013年第1期，第35—42页。
7. 冈萨雷斯-托雷斯："精神分析与神经科学"，第36页。
8. 冈萨雷斯-托雷斯："精神分析与神经科学"，第37页。
9. 冈萨雷斯-托雷斯："精神分析与神经科学"，第37页。
10. 乔治·弗兰克："回应'西格蒙德对于21世纪的意义'一文"，《精神分析心理学》第25卷，2008年第2期，第375—379页。
11. 冈萨雷斯-托雷斯："精神分析与神经科学"，第38页。
12. 乔治·弗兰克："回应"，第376页。
13. 尼玛·巴西里："弗洛伊德与脑物质：论神经精神分析的重组"，《批判性探索》第40卷，2013年第1期，第83—108页。

12 未来展望

要点 🗝

- 《梦的解析》未来最大的潜能在于其对神经精神分析学＊的支撑［神经精神分析学是指对精神分析学与神经科学＊的交叉研究；神经科学是对大脑与神经系统的研究］。
- 《梦的解析》将在心理学与精神病学的历史中继续扮演重要的角色。
- 《梦的解析》仍然是心理学与精神病学中最核心和最有影响力的文本之一。

潜力

西格蒙德·弗洛伊德的《梦的解析》在心理学和精神病学领域开创了独树一帜的思想流派。从这个意义上讲，它已经实现了它的全部潜能。精神分析学曾有几十年是心理学的主导理论，或许它在美国的影响最为明显。不过，自20世纪70年代以来，精神分析学家的人数与影响急剧下滑。当不同类型的心理疾病可以用药理学方法进行治疗时，人们的兴趣转向了更具有实证性形式的治疗方法（药理学是指关于药物使用的科学）。[1] 精神分析学能否在未来重获其巨大的影响力，现在还仍然难以想象。不过，《梦的解析》毫无疑问将在两人领域保持其经典佳作的地位。专业医疗人士与非专业读者将因为其重要的历史价值而继续对它进行研读。

据悉，神经科学家们为了更好地理解人的心理，已经对弗洛伊德精神分析学的一些思想产生了新的兴趣。正因为如此，越来越多的人呼吁将这两个学科融合成一个新的研究领域，即神经精神分析学。这个学科已经获得了重要的支持。诺贝尔奖获得者、神经精神病学家埃里克·坎德尔*说，"我希望在开创崭新的、令人信服的心理与心理疾病理论时，如果能融合认知神经科学，那么，精神分析学将重获其智慧的能量。"² 这样的融合可能代表了弗洛伊德思想在未来的最大潜能。

> "我的目的……在于提出一种能让精神分析学重焕生机的方法，也就是与生物学，特别是与认知神经学建立一种更加紧密的普遍关系。"
> ——埃里克·R. 坎德尔："生物学与精神分析学的未来：重访精神病学的新知识框架"

未来方向

在1999年的一篇论文中，坎德尔为神经精神分析学搭建了一个框架。"精神分析学的核心思想是，"坎德尔写道，"我们对我们精神生活的大部分内容不甚了解。"³ 神经科学家们已经对外显（有意识的）记忆*与内隐或程序性记忆*作出区分。"程序性记忆"此处是指无意识的记忆，例如与完成某个过程（如任务）的技能紧密相关的记忆。

神经精神分析学可以通过借助精神分析学的心理观来描绘这些差异。坎德尔指出，"无意识心理过程早期研究的局限性之一是没有对论点直接进行验证的方法。"不过，"当下生物学——以其能够描绘精神过程的能力，以及对患者程序性记忆构成不同损害的器官的研究能力——能够做出一个重要贡献就是将无意识精神过程的研究基础从间接的推断改变为直接的观察。"[4]

坎德尔注意到,弗洛伊德所使用过的一些方法,如自由联想法,同样可以借用到生物学研究领域。"在生物学中,我们在理解联想如何在程序性记忆的形成方面已经有了良好的开端……只要程序性知识的各个层面与意义产生的瞬间相互关联,那么这些生物学的研究视角在理解程序性无意识的内容方面将证明是有用的。"[5] 他看到精神分析学与神经科学之间进一步融合的潜能。这些融合包括思维与精神病理学之间的联系,童年经历在精神病理学中的作用,以及人脑中被称之为前额皮质*、与更高认知技能相联系的那部分在前意识思维中所扮演的角色。"前意识"此处是指位于无意识与意识之间的那部分心理内容;心理中没有遭受压抑的记忆部分,在没有被召唤到意识之前都存储在这里。

小结

西格蒙德·弗洛伊德的《梦的解析》仍然是心理学和精神病学领域的一个经典文本。当然,我们不能夸大弗洛伊德的影响、这本著作的影响,以及精神分析学派的崛起对这些研究领域的影响。坎德尔对此作出了很好的总结。他说,"在20世纪上半叶,精神分析学彻底改变了我们对精神生活的认识。它就无意识心理过程提出了一系列杰出而崭新的洞见。"[6] 学术界将弗洛伊德的天才与英国进化论学者查尔斯·达尔文、德国物理学家阿尔伯特·爱因斯坦相提并论。尽管弗洛伊德是一位多产作家,但是学界一直将《梦的解析》视作他最重要的作品。[7] 任何读者只要对心理学与精神病学的历史、对神经科学,或者对弗洛伊德影响人类文化的方式感兴趣,都会从阅读《梦的解析》中获益匪浅。神经科学家们重新燃起了他们对精神分析理论的兴趣，这本著作在未来仍然会显示出重要意义。

我们始终能从这本书中获得新颖的洞见——事实上，在最近几十年中，关于无意识心理的许多概念已经发生了变化。但《梦的解析》提出了人类历史上最著名、最有影响力的心理理论之一。尽管它的重要性在弗洛伊德之后可能发生了改变，但是其学术价值是毋庸置疑的。

---

1. 保尔·R. 麦克休："弗洛伊德之死与精神病学之重生"，《标准周刊》，登录日期 2016 年 1 月 10 日，http://www.weeklystandard.com/article/12226。
2. 埃里克·R·坎德尔："生物学与精神分析学的未来：重访精神病学的新知识框架"，《美国精神病学学刊》第 156 卷，1999 年第 4 期，第 505—524 页。
3. 坎德尔："生物学与精神分析学的未来"，第 508 页。
4. 坎德尔："生物学与精神分析学的未来"，第 510 页。
5. 坎德尔："生物学与精神分析学的未来"，第 510 页。
6. 坎德尔："生物学与精神分析学的未来"，第 505 页。
7. 帕特里夏·凯切尔：《弗洛伊德的梦：跨学科的心理科学全景》，麻省剑桥：麻省理工学院出版社，1992 年。

术语表

1. **分析心理学**：弗洛伊德的学生卡尔·荣格创立的另一支精神分析学思想流派。

2. **反犹主义**：对犹太人的偏见。

3. **原型**：在本书语境中，是指集体无意识中出现的意象与模型（精神分析学家卡尔·荣格创造的术语，是指特定人群或社区共享的部分无意识心理）。

4. **生物医学**：医学中对生物学原理的运用。

5. **集体无意识**：荣格创造的术语，指所有人共享的那部分无意识。

6. **凝缩**：在本书语境中，指将几种欲念压缩到显梦内容中的一个元素。

7. **认知神经学**：神经学的一个分支，对与行为与思维过程中人脑的生物功能的研究。

8. **认知心理学**：对与思维相关的人类心理与行为的研究。

9. **良知**：对是非对错的感知。

10. **意识**：人的感知状态。

11. **防御（自我防御机制）**：应对焦虑的无意识技巧。

12. **妄想**：某种错误的信念。

13. **置换**：在本书语境中，指隐梦内容（隐蔽而"真实"）与显梦内容（可描述）之间的相互替换。

14. **梦的解析**：在本书语境中，指运用自由联想法确定梦的真实含义的过程。

15. **自我**：根据弗洛伊德的定义，指活动在本我与超我区间的那部分意识。

16. **启蒙运动**：17和18世纪（西方）文化、智性与哲学运动，强调通过教育、科学、个人主义和理性来达到社会与个人进步。

17. **经验主义的**：以观察和实证为依据。

18. **性快感区**：根据弗洛伊德理论，是指性能量集中的身体部位。这个区域随着人的成长而出现在不同部位。

19. **显性记忆**：人能够清醒意识到和回想到的那些记忆。

20. **病因学**：研究发病的原因。

21. **固恋**：在性心理的发展过程中，对性快感区的固执依恋而没能成功消除。

22. **自由联想法**：心理学实验方法。一个人（比如治疗师）说出一个词，另一个人（病人）立刻用另一个词作出回应。这两个词之间可能没有明显的联系。

23. **弗洛伊德式口误**：言语失误，被认为能显示个体无意识欲望的相关信息。

24. **人文学科**：学术界研究人类文化的学科。

25. **催眠**：让一个人在接受暗示后经受某种意识状态被改变的心理过程。

26. **歇斯底里**：在本书语境中，指由极端情绪和异常感观式生理功能导致的神经失常。

27. **本我**：无意识欲望和心理能力的储存地。

28. **惯性**：在本书语境中，指极度兴奋在神经系统中的消除。

29. **婴幼儿的**：婴幼儿时期的。

30. **隐梦**：梦的真实含义。

31. **显梦**：梦的故事线。

32. **唯物主义**：在本书语境中，是指所有心理过程源自大脑功能的假说。

33. **能动输出**：从神经系统发往肌肉的信号，致使身体运动或行动。

34. **自然科学**：研究自然界的学科领域。

35. **纳粹德国（1933—1945）**：阿道夫·希特勒与他的纳粹党统治德国和在第二次世界大战时统治其他几个国家的历史时期。纳粹强烈反犹，最终囚禁并杀戮了6百万犹太人、同性恋者以及他们视为"不良分子"的人。

36. **新弗洛伊德派**：是指那些受到弗洛伊德影响但是对弗洛伊德的精神分析学某些论断持反对态度的理论家和治疗师。

37. **神经元活动**：大脑神经细胞区域的活动。

38. **神经病专家**：专门治疗神经系统与大脑疾病的医生。

39. **神经病理学**：研究神经系统疾病的医学分支。

40. **神经生理学**：研究神经系统生理功能的学科。

41. **神经精神病学**：研究神经系统在精神疾病中的作用的学科。

42. **神经精神分析学**：这个术语是指心理分析与神经科学方法合二为一的研究。

43. **神经科学**：专门研究神经系统结构与功能的交叉研究领域。

44. **神经症（精神神经病）**：焦虑型的精神疾病。

45. **神经症患者**：在特定心理状态下出现心理焦虑症状的病人。

46. **迷恋**：对某个念头的强烈痴迷，经常与焦虑密切相关。

47. **俄狄浦斯情结**：弗洛伊德提出的观点，认为人在正常的心理发展过程中，会出现（最终会压抑）对对立性别的父或母的性冲动，同时将同性别的母或父视作威胁。

48. **病理学的**：与疾病相关的。

49. **人格结构**：个体思想与行为的性格模型。

50. **恐惧症**：强烈、非理性的害怕。

51. **前意识**：处在无意识与意识之间的那部分心理区域。各种没有被压抑的记忆在被召回到意识之前所储存的区域。

52. **前额皮质**：与更高的认知功能相关的大脑区域。

53. **程序记忆**：如何处理任务的记忆；一个人经常不需要有意识地回想的记忆。

54. **投射**：弗洛伊德式的自我防御机制，个体通过这个机制将自身焦虑不安的思想归咎到别人身上。

55. **心理**：人的内心。

56. **精神病学**：治疗精神疾病的医学分支。

57. **精神性的**：在本书语境中，是指精神过程。

58. **精神分析学**：一个思想流派，强调无意识心理在意识行为中的作用；也是一种治疗方法，通过采用不同技巧试图洞悉无意识心理。

59. **精神动力学**：一个思想流派，强调无意识心理对行为的形塑。

60. **心理学**：研究心理过程与行为的科学。

61. **精神神经病（亦称神经官能症）**：焦虑型的精神疾病。

62. **精神病理学**：研究心理疾病的学科。

63. **理性主义**：重视理性与逻辑的哲学思想。

64. **合理化**：弗洛伊德式的自我防御机制，个体试图寻找托词来自证自身的行为。

65. **反射弧**：弗洛伊德时代通行的概念，认为人的行为是对感知输入或感知所作出的反射性回应。

66. **反射循环**：感知输入与运动输出的神经元循环。

67. **表征化**：在本书语境中，是指心理在显梦中呈现隐秘思想的方式。

68. **压抑**：弗洛伊德式的自我防御机制，是指将焦虑不适的念头和记忆从意识感觉区域清理出去。

69. **润饰**：无意识提供给显梦的连贯性叙事。

70. **感知输入**：感知接收器接受到信息，然后将信息传送到大脑。

71. **性心理发展的阶段论**：弗洛伊德提出的观点，认为在性心理的发展中，个体的性快感区在特定时期集中在身体某个部位，直至性冲突消除。无论消除还是没有被消除，这些性冲突都会对个体的人格结构带来长远的影响。

72. **超我**：含有良知感的那部分意识心理。

73. **心理理论**：弗洛伊德尝试阐释他人心理过程的思想。

74. **无意识心理**：处于感觉意识之外的那部分心理意识。

75. **维多利亚时代**：19世纪至20世纪早期的特定历史时期（大致是指大不列颠维多利亚女王在位期间）。此处非常重要的是，维多利亚时代的人对性有着强烈的道德关注。

76. **梦遗**：睡梦中遗精现象。

人名表

1. 阿尔弗雷德·阿德勒（1870—1937），奥地利医生与精神病学家。作为弗洛伊德早前的追随者，阿德勒继续前行，创立了个体心理学。

2. 约瑟夫·布洛伊尔（1842—1925），奥地利医生，弗洛伊德的导师，与他合作撰写了《癔症研究》。

3. A. A. 布里尔（1847—1948）：奥地利精神病专家，最早将《梦的解析》翻译成英文。

4. 让-马丁·沙可（1825—1893），法国神经病学家，弗洛伊德的导师。沙可经常被视作神经病学之父。

5. 查尔斯·达尔文（1809—1882），英国生物学家，因提出进化论思想而闻名于世。

6. 阿尔伯特·爱因斯坦（1879—1955），出生于德国的物理学家，因提出广义相对论而闻名于世。

7. 桑多尔·费伦齐（1873—1933），匈牙利精神分析学家，弗洛伊德晚年时期的一位同事。他提出了与弗洛伊德不同的观点，认为精神分析学家不应该是相关过程中的被动参与者，而应该发挥更加主动的作用。

8. 约翰·约瑟夫·加斯纳（1727—1779），奥地利治病术士与驱魔术士。加斯纳与弗朗兹·安东·麦斯麦尔的对峙开启了精神病学的新阶段。

9. 彼得·盖伊（1923—2015），德裔美国历史学家，主要研究欧洲文化与知识史。

10. 伊俄卡斯忒王后，古希腊神话中的人物，拉伊俄斯王的王后，俄狄浦斯王的母亲。

11. 厄尼斯特·琼斯（1879—1958），英国精神分析学家，弗洛伊德传记的作者。

12. **卡尔·荣格**（1875—1961），瑞士精神病学家，弗洛伊德最出色的同事之一。荣格后来在精神病学领域创立了心理分析学。

13. **埃里克·坎德尔**（1929年生），奥地利裔美国神经精神病学家，因为从事记忆的生物学基础研究于2000年获得诺贝尔奖。

14. **拉伊俄斯王**，古希腊神话中的人物，俄狄浦斯王的父亲。

15. **卡尔·马克思**（1818—1883），德国哲学家，因为其著作《共产党宣言》和《资本论》而驰誉天下。

16. **弗朗兹·安东·麦斯麦尔**（1734—1815），德国精神病科医生，因提出动物磁性论而闻名于世。

17. **西奥多·梅涅特**（1833—1892），德裔奥地利神经病理学家，曾经是弗洛伊德早年在维也纳医院精神病科工作时的指导人。

18. **安娜·欧**，约瑟夫·布洛伊尔最著名的病人的假名。她的案例研究成为他与弗洛伊德合著的推手。

19. **俄狄浦斯王**，古希腊神话中的人物。他在不知情的情况下杀死了父亲拉伊俄斯王，娶了母亲伊俄卡斯忒王后。

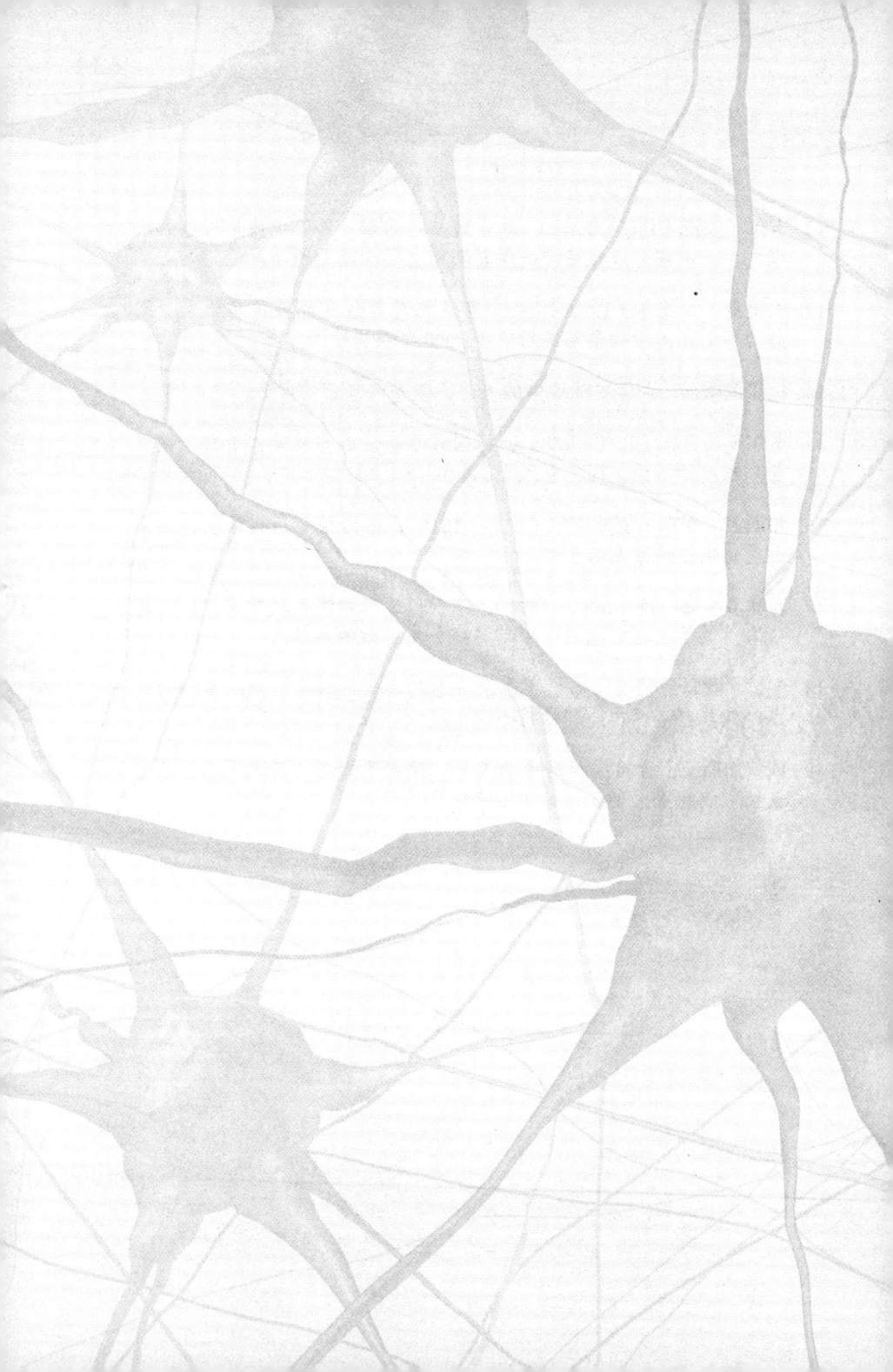

WAYS IN TO THE TEXT

KEY POINTS

- Sigmund Freud was an Austrian neurologist* (a specialist in the nervous system and brain) and the founder of the therapeutic and theoretical approach to the unconscious mind* known as psychoanalysis.*

- In *The Interpretation of Dreams*, Freud argues that we can understand the wishes of the unconscious mind by analyzing and interpreting the meaning of dreams.

- Scholars consider *The Interpretation of Dreams* to be Freud's most important work; it has greatly influenced the fields of psychology* (the study of the human mind and behavior) and psychiatry* (the treatment of disorders of the mind).

Who Was Sigmund Freud?

Born in 1856, Sigmund Freud lived most of his life in Vienna, Austria. After receiving his medical degree from the University of Vienna in 1881, he worked in the psychiatric unit at a local hospital. This gave him hands-on experience of working with individuals who suffered from mental illnesses.

In 1885, Freud studied under the famed French neurologist Jean-Martin Charcot.* While working with Charcot, Freud realized that many psychological disorders trace back to traumatic events earlier in a patient's life. These events get stored in the unconscious mind. In 1893, the Austrian physician Josef Breuer* told Freud about his work with a patient who had some unusual physical symptoms with no obvious physical causes. Rather than treat the patient medically, Breuer began to engage in a form of therapy in which the patient talked about his experiences. Over time, Breuer found that this "talking cure" improved his patient's symptoms. In 1895, Freud and Breuer described this new therapeutic approach in a book called *Studies on Hysteria*.

Freud's experiences with Charcot and Breuer inspired his ideas about the role the unconscious mind plays in determining behavior. He began to view dreams as one way the unconscious mind communicates its desires. Freud's 1899 masterwork *The Interpretation of Dreams* became wildly popular and helped establish the psychoanalytic perspective in psychology and psychiatry. The work has since been published around the world and remains in print more than a century later.

What Does *The Interpretation of Dreams* Say?

Before Freud, most people believed that dreams had no purpose and no meaning. Freud made the radical argument that dreams, created by the unconscious mind, in fact contain a great deal of meaning.

In *The Interpretation of Dreams*, Freud argues that dreams contain meaningful information about the dreamer's unfulfilled, unconscious wishes. Each dream, he claims, is a message in disguise; one must interpret its meaning through dream analysis:* the process of talking through the associations the dreamer might hold for what the dream contains.

Freud begins *The Interpretation of Dreams* by reviewing the existing literature on dreams. He discusses the ancient idea that dreams were the work of spirits or demons. He also examines the more contemporary notion that dreams represent extensions of the waking state or neural activity* (electrical and chemical processes at the level of cells) in the brain's sensory regions.

Freud demonstrates how to analyze and interpret a dream. He describes engaging the patient in free association* as he considers each aspect of the dream. In free association, one person (the therapist, for instance) says a word and the other (the patient) responds immediately with another word. The words may have no any obvious connection, but the therapist can use the sequence to interpret the workings of the patient's unconscious mind.

Freud also showed that all dreams represent the fulfillment of some unconscious wish. Our conscious minds censor these wishes in some way, so they manifest in our dreams. The dream story, or "manifest content,"* hides the dream's true meaning, or "latent" content.* Freud describes several ways in which the mind transforms wishes into dreams. This allows the reader to better engage in dream analysis.

The unconscious transforms wishes into dreams if they would be too uncomfortable to acknowledge in an uncensored form.According to Freud, many of these wishes represent sexual impulses. These sexual impulses often get established early in life, as a child interacts with his or her parents. Freud believed that it was a normal part of human development for children to direct sexual and aggressive impulses at their parents. In his view, the difference between healthy people and those suffering from psychological problems is that the latter group exaggerated these normal impulses.

Freud describes his rough theory of mind* in terms of reflex circuits*—simple connections in which the mind receives sensory information and the body responds by acting in some way. This biological conception of the mind, common in his day, reflected Freud's training and background as a physician. He saw the mind as consisting of two parts: unconscious and conscious. With no direct access to the conscious mind, the unconscious mind relies on dreams to convey its wants and desires into the realm of consciousness.* This idea served as a foundation for much of Freud's later work and distinguished him from many of his peers.

Why Does *The Interpretation of Dreams* Matter?

Scholars generally consider *The Interpretation of Dreams* to be Freud's most significant work. In it, he lays the foundation for many of his future ideas. It helped established the psychoanalytic school of thought in the fields of psychology and psychiatry. It also offers a technique for treating psychological problems. The psychoanalytic school emphasizes the role that the unconscious mind plays in motivating conscious behavior. Freud also developed a "stage theory" of psychosexual development,* which used principles of psychoanalysis to explain (very roughly) normal human development in terms of stages differing in the source of sexual pleasure. Freud believed that an adult's personality* traits reflect the resolution of sexual impulses as that person passed through the various stages of development.

The Interpretation of Dreams and Freud's psychoanalytic

perspective played a great part in shaping the disciplines of both psychology and psychiatry. The work helped people think about the unconscious mind in entirely new ways (the idea that dreams carry meaningful information from the unconscious, for example). As Freud's ideas inspired followers, the psychoanalytic perspective grew more influential. Some of the people inspired by his work held different opinions of the exact nature of the unconscious mind and about the importance of sexual impulses.As a result of these disagreements, some followers established their own perspectives; with ideas clearly based on Freud's original conception, these people are called "neo-Freudians."*

The criticisms the text inspired may be as important as the text itself. Almost from the outset, some of Freud's contemporaries dismissed his ideas as flawed. Critics saw them as overly anecdotal (specific examples that cannot be considered truly representative), since they arose from the work Freud had done with his patients and from his own self-analysis. Beyond being anecdotal, they were not testable. Science demands that ideas be repeatable and verifiable—but no one can observe the unconscious mind. Although the work may have seemed unscientific, Freud clearly valued science; this becomes clear on reading the text. Still, advances in science may open the door to exploring old ideas in new ways, enabling us to better verify the conclusions Freud drew from his and his patients' experiences.

SECTION 1
INFLUENCES

MODULE 1
THE AUTHOR AND THE HISTORICAL CONTEXT

KEY POINTS

* *The Interpretation of Dreams* lays the groundwork for Freud's theories, which went on to become very influential and changed psychology*—the study of the human mind and behavior—forever.
* Freud's analysis of his own dreams and the death of his father served as a springboard for *The Interpretation of Dreams*.
* The morality of the Victorian era*—the period bridging the nineteenth and early twentieth centuries marked by the reign of the British monarch Queen Victoria—dominated Europe during much of Freud's life, playing a large part in shaping his theories.

Why Read This Text?

The Interpretation of Dreams is Sigmund Freud's most famous work. Published in 1899, it cemented his reputation as the father of the theoretical and therapeutic model of the psychoanalytic* movement that dominated the field of psychology for several decades.¹ Although many of his original ideas have proven problematic, Freud has undoubtedly been one of the most influential figures in the history of psychology. Indeed, as one scholar noted, "Freud has been ranked along with [the political philosopher] Karl Marx,* [the foundational evolutionary theorist] Charles Darwin,* and [the physicist] Albert Einstein* as the several geniuses who have put their imprints so very decisively upon the ways in which we have to come understand the world in which we live and our place in it."² So influential was Freud that "our very language was... saturated with 'Freudian' meaning, slips* of the tongue, repression,* projection,* rationalization,* defenses,* and so forth"—all terms taken from Freudian theory, relating to unconscious speech, the unconscious mind* and behavior, and the therapeutic process.³

The Austrian psychiatrist* A. A. Brill* produced the first English translation of the book in 1913. He called *The Interpretation of Dreams* "the author's greatest and most important work."⁴ In the book, Freud established the foundation on which he built his theories of the human mind and personality.* The book also guided other professionals who followed in Freud's footsteps as psychoanalysts. It represents an important text in the history of psychoanalysis and psychology, offering the reader insight into Freud's ideas about the important role dreams play in conveying information about the unconscious mind.

> *"The dream proves to be the first link in a chain of abnormal psychic* structures whose other links, the hysterical phobia,* the obsession,* and the delusion* must, for practical reasons, claim the interest of the physician."*
>
> —— Sigmund Freud, *The Interpretation of Dreams*

Author's Life

Born in 1856, Sigmund Freud spent most of his life in his home city of Vienna, Austria. He studied medicine at the University of Vienna, receiving his medical degree in 1881. The following year, he became a clinical assistant at the Vienna General Hospital, where he worked in the psychiatric clinic under neuropathologist* Theodor Meynert* (neuropathology is the study of disorders of the nervous system and brain). In 1885, Freud received a scholarship to study with the Frenchman considered "the father of neurology,"* Jean-Martin Charcot.* Charcot taught Freud about the symptoms of hysteria* (a neurotic* disorder characterized by extreme emotion) and how to treat them using hypnosis.*[5]

Returning to Vienna, Freud opened a private practice specializing in treating patients with nervous disorders. In 1893, he began to collaborate with the Austrian physician Josef Breuer.* The two men cowrote the book *Studies on Hysteria*,[6] (1895) which introduced free association*—a method in which patients are asked to offer immediate responses to words given to them by the psychoanalyst— as a therapeutic technique to treat hysteria. During this period, Freud also began to analyze and interpret some of his own dreams. He documented the experience, along with his work with patients and the death of his father, in *The Interpretation of Dreams*. In his preface to the book's second edition, Freud described it as "a part of my selfanalysis, a reaction to the death of my father—that is, to the most significant event, the deepest loss, in the life of a man."[7]

After the first edition of *The Interpretation of Dreams* was published in 1899, Freud established what came to be known as the Vienna Psychoanalytic Society. He published several more books over the course of his career, elaborating on the ideas introduced in *Interpretation*. As a secular Jew, Freud was affected by the anti-Semitic* sentiments emanating from nearby Nazi Germany* (anti-

Semitism is hostility to Jewish people). Germany annexed Austria in 1938, and Freud and his family fled to England a few months later. He died in London on September 23, 1939.[8]

Author's Background

The Swiss psychiatrist Carl Jung,* who studied with Freud, said that "the historical conditions which preceded Freud and formed his groundwork made a phenomenon like himself necessary."[9] Indeed, the times in which Freud lived played a large role in the formation of some of his ideas. One of the strongest of these influences was the Victorian morality so pronounced in late nineteenth-century Europe. Victorians exhibited an "intense moral preoccupation with sexuality... Sexuality not only needed to be regulated by personal morality, or by the vigilance of the family, but, since it could affect entire populations, was a political and social concern."[10]

During this period, society also became increasingly interested in "scientific materialism* [here the assumption that all mental processes are a function of the brain] and [the logically-oriented philosophical position of] rationalism,"* wrote Jung. "This is the matrix out of which Freud grew, and it is the mental characteristics of this matrix which have shaped him along foreordained lines."[11]

Freud believed that the unconscious mind influences our behavior. We can tap into the unfulfilled wishes of the unconscious by analyzing our dreams. Given the sexual repression of the Victorian era, we should not be surprised that Freud believed the unconscious mind played a central role in repressing sexual attitudes and urges.

1. B. M. Thorne and T. B. Henley, *Connections in the History and Systems of Psychology* (Boston, MA: Houghton Mifflin Company, 2005).
2. Robert S. Wallerstein, "The Relevance of Freud's Psychoanalysis in the 21st Century: Its Science and Its Research", *Psychoanalytic Psychology* 23, no. 2 (2006): 302–326.
3. Wallerstein, "The Relevance of Freud's Psychoanalysis", 303.
4. Sigmund Freud, *The Interpretation of Dreams*, translated by A. A. Brill, with an introduction and notes by Daniel T. O'Hara and Gina Masucci MacKenzie (New York: Barnes & Noble Books, 2005), 9.
5. Freud, *The Interpretation of Dreams*, xi–xii.
6. Sigmund Freud and Josef Breuer, *Studies on Hysteria*, trans. James Strachey (London: Hogarth Press, 1955).
7. Freud, *The Interpretation of Dreams*, 6.
8. Freud, *The Interpretation of Dreams*, xiii–xix.
9. C. G. Jung, "Sigmund Freud in His Historical Setting", *Journal of Personality* 1, no. 1 (1932), 48–55.
10. "Historical Context for the Writings of Sigmund Freud", *Columbia College: The Core Curriculum*, accessed January 2, 2016, http://www.college.columbia.edu/core/content/writings-sigmund-freud/context.
11. Jung, "Sigmund Freud in His Historical Setting", 49.

MODULE 2
ACADEMIC CONTEXT

KEY POINTS
- While many view Sigmund Freud as the father of the therapeutic methods and theories of psychoanalysis,* the foundation for the psychoanalytic perspective was actually laid by the German physician and hypnotist* Franz Anton Mesmer* some 80 years before Freud's birth.
- By the time Freud entered the field, neuropsychiatry*—the field examining the role of the nervous system in disorders of the mind—had taken a primarily biological perspective in explaining psychological disorders.
- Freud's work with both the physicians Jean-Martin Charcot* and Josef Breuer* greatly influenced him in establishing the perspective of psychoanalysis.

The Work in its Context

Many consider *The Interpretation of Dreams* to be Sigmund Freud's most important work. It established psychoanalysis (sometimes called psychodynamic* therapy—a school of thought that emphasizes the role of unconscious forces in shaping behavior) as a dominant perspective in the fields of psychology* and psychiatry.*

Freud asserted that the unconscious mind* profoundly influences our conscious behavior and personality.*This new perspective created fundamental shifts in the disciplines that treated psychological disorders. Indeed, they would go on to rock the culture as a whole. But Freud did not introduce the concept of the unconscious mind. Some have argued that "a clash between the physician Franz Anton Mesmer and the exorcist Johann Joseph Gassner"*[1] in 1775 laid the foundations for psychoanalysis and the concept of the unconscious mind. ("Exorcism" is the ritual expulsion of malign forces such as demons from the troubled body.)

Gassner, a very popular healer, used religious techniques to rid people of various maladies. By 1775, as the rationality of the intellectual and social movement known as the Enlightenment* became more prominent in Europe, Gassner and his mystical cures came under suspicion. Mesmer provided an alternate perspective. Being a trained physician, he took a more scientific approach to the treatment of nervous disorders. Ultimately, the medical establishment rejected Mesmer's ideas. However, his opposition to spiritual modes of healing set a change into motion, and ultimately, clinical professionals adopted the psychoanalytic approach.[2]

> *"With Freud begins the era of the newer dynamic schools, with their official doctrine, their rigid organization, their specialized journals, their closed membership, and the prolonged initiation imposed upon their members."*
> ——Henri F. Ellenberger, *The Discovery of the Unconscious: The History and Evolution of Dynamic Psychiatry*

Overview of the Field

The treatment of mental illness had already changed quite a bit by the time Sigmund Freud entered the profession. In the early days, people believed that mentally ill people were possessed by demons, so exorcists like Gassner treated them by spiritual means. By the late nineteenth century, most professionals in the field took a much more biological and systematic approach to understanding these disorders. They focused on tangible (that is, concrete), directly observed aspects of the person, such as their physical state. The shift toward science had been so complete that Freud felt compelled to write that his focus on the unconscious through dream analysis* had not "overstepped the bounds of neuropathological* interest"[3] ("neuropathology" here indicating disorders of the nervous system and brain). Referencing phobias* (irrational fears) and delusion* (false beliefs), he added, "one that cannot explain the origin of the dream pictures will strive in vain to understand the phobias, obsessive* and delusional ideas, and likewise their therapeutic importance."[4] Clearly, Freud worried that the establishment had largely discounted the influence of psychological factors on behavior.

However, some in the field shared Freud's views. They believed that both the unconscious mind and psychological factors influenced their patients' symptoms. Freud's work with these like-minded individuals, such as Charcot, in the formative years of his career helped him shape the ideas in *The Interpretation of Dreams*.

Academic Influences

A scholarship allowed Freud to study in Paris with the French physician Jean-Martin Charcot, whom many consider to be the father of neurology.* Charcot specialized in treating people with unusual symptoms that had no apparent physical cause. Over the course of his career, Charcot came to believe that his patients were experiencing "a form of hysteria* which had been induced by their emotional reaction to a traumatic accident in their past."[5] During his time with Charcot, Freud concluded that neuroses*—mental illnesses involving anxiety—originating from the unconscious mind caused these unusual symptoms.

Freud's collaborations with the respected Austrian physician Josef Breuer also deeply influenced him. Breuer helped Freud establish his own medical practice and in many ways served as one of his mentors. Breuer told Freud about one of his most interesting patients, a woman who came to be known as Anna O.* They would recount her story in the book they coauthored in 1895 entitled *Studies on Hysteria*.[6] Anna's bizarre symptoms included unexplained coughing, paralysis of one side of her body, and even vivid hallucinations. Ultimately, Breuer diagnosed her with hysteria. But he found that when he discussed the hallucinations with Anna, her symptoms faded. This talking cure intrigued Freud. Combining the ideas of trauma-induced hysteria with Breuer's form of therapy, Freud developed his unique approach. After Freud, free association* (the process in which a patient immediately responds to words, without self-censorship) and dream analysis (the process of discovering what the unconscious mind is communicating in dreams) would become cornerstones of psychoanalysis.[7]

---

1. Henri F. Ellenberger, *The Discovery of the Unconscious: The History and Evolution of Dynamic Psychiatry* (New York: Basic Books, 2008), 53.

2. Ellenberger, *The Discovery of the Unconscious*, 53–57.
3. Sigmund Freud, *The Interpretation of Dreams*, translated by A. A. Brill, with an introduction and notes by Daniel T. O'Hara and Gina Masucci MacKenzie (New York: Barnes & Noble Books, 2005), 3.
4. Freud, *The Interpretation of Dreams*, 3.
5. Richard Webster, "Freud, Charcot, and Hysteria: Lost in the Labyrinth", accessed January 3, 2016, http://www.richardwebster.net/freudandcharcot.html.
6. Sigmund Freud and Josef Breuer, *Studies on Hysteria*, trans. James Strachey (London: Hogarth, 1955).
7. Webster, "Freud, Charcot, and Hysteria: Lost in the Labyrinth".

MODULE 3
THE PROBLEM

KEY POINTS
- Freud and his contemporaries wanted to better understand the psychological causes of some mental illnesses.
- Most of Freud's colleagues sought physiological explanations for these disturbances. In doing so, they largely ignored the unconscious mind* and dream analysis.*
- Freud believed that the unconscious mind played an important role in causing many mental illnesses. He also held that dreams enabled the unconscious mind to reveal unfulfilled wishes.

Core Question

In *The Interpretation of Dreams*, Sigmund Freud presented his belief that dreams provide insight into the unconscious mind, and that analyzing them could help treat mental illnesses. At the time Freud wrote the book, the medical establishment paid little attention to dreams or the unconscious mind. They focused instead on trying to determine the physical causes of mental illnesses. Back then there were few technological resources to map out the nervous system. Instead, professionals proposed that behavior was a reflex circuit: * circuits, constructed in the brain, that involved sensory input* (things such as sight and hearing) and motor output* (physical action). Freud noted that "the reflex arc* remains the model for every psychic* activity"[1] so the theory of mind* he presented in *The Interpretation of Dreams* followed this essential organization.

In Freud's view, the core question of the day involved determining what role, if any, the unconscious mind plays in affecting conscious behavior. Assuming his emphasis on the unconscious was correct, in *The Interpretation of Dreams* Freud outlined a procedure other clinicians could use to analyze and interpret dreams. By couching these ideas in scientific language, Freud hoped that the book would appeal to a broad range of colleagues across the scientific and spiritual communities.

> *"It is hard to deny the extent to which Freud's energetic theory of reflex diverged from the dominant nineteenth century conceptions."*
> —— Nima Bassiri, "Freud and the Matter of the Brain"

The Participants

Freud was convinced that some types of psychological disturbances were psychic rather than physical in nature. His work with both the French neurologist* Jean-Martin Charcot* and the Austrian physician Josef Breuer* only strengthened this belief, but many of his contemporaries disagreed. Freud had trained in medicine and developed his theories of psychoanalysis* from an increased commitment to a scientific understanding of the mind. But "Freud openly broke with official medicine"[2] by taking the position that mental illness could cause physical symptoms. Some critics considered the dream to be a simple "reaction to the stimulus causing a disturbance of sleep."[3] As such, the established medical position held that dreams served no real purpose. Without a purpose, they could offer no special understanding of either the human mind or the symptoms of hysteria.*

But Freud believed "the dream has a meaning, albeit a hidden one; that it is intended as a substitute for some other thought process, and that it is only a question of revealing this substitute correctly in order to reach the hidden signification of the dream."[4] He used *The Interpretation of Dreams* to communicate his ideas with his colleagues. Ultimately, he believed science would turn to his approach in treating individuals suffering from physical symptoms when doctors could identify no somatic—that is, body-based—abnormalities.

The Contemporary Debate

The idea of behavior as a reflex had become quite popular by the time Freud wrote *The Interpretation of Dreams*. Indeed, Freud himself had described behavior in terms of a reflex before he wrote the book. As one writer noted, "Freud defined reflex action as a tendency towards a state of inertia."*[5] Freud used "inertia" in a specific sense; it indicated "the tendency of the nervous system—and thus the psyche* [the mind] to discharge excessive buildups of excitation."[6] Freud and his contemporaries believed that people derived pleasure from this discharge of excess excitation. If this excitation was allowed to accumulate, however, it led to aversion—a desire to avoid something.

While Freud used the popular model of the time to introduce his ideas, his concept of reflex actually differed significantly from the more prominent, generally accepted ideas. Specifically, Freud's focus on the role of the unconscious mind bore little similarity to the connections between sensory and motor systems generally described as "reflexes." Critics noted the differences and attacked Freud's ideas about the mind and how dreams help us understand its operations. Despite the criticism, Freud sometimes felt ignored and unappreciated by the larger psychiatric* community. Indeed, in the preface to the second edition of his book, he remarked that, "the behavior of the scientific critics could only justify the expectation that this work of mine was destined to be buried in oblivion."[7]

---

1. Sigmund Freud, *The Interpretation of Dreams*, translated by A. A. Brill, with an introduction and notes by Daniel T. O'Hara and Gina Masucci MacKenzie (New York: Barnes & Noble Books, 2005), 425.
2. Henri F. Ellenberger, *The Discovery of the Unconscious: The History and Evolution of Dynamic Psychiatry* (New York: Basic Books, 2008), 418.
3. Freud, *The Interpretation of Dreams*, 73.
4. Freud, *The Interpretation of Dreams*, 89.
5. Nima Bassiri, "Freud and the Matter of the Brain: On the Rearrangements of Neuropsychoanalysis", *Critical Inquiry* 40, no. 1 (2013): 83–108.
6. Bassiri, "Freud and the Matter of the Brain", 91.
7. Freud, *The Interpretation of Dreams*, 5.

MODULE 4
THE AUTHOR'S CONTRIBUTION

KEY POINTS

- Freud believed that the psyche,* or the mind, actively constructed dreams to communicate information about the dreamer's unfulfilled wishes or desires.
- Freud's focus on the unconscious mind* offered innovative, alternative explanations to the physical causes of psychological disorders.
- In *The Interpretation of Dreams*, Freud creatively synthesizes and extends the work he did with both the physicians Jean-Martin Charcot* and Josef Breuer.*

Author's Aims

In writing *The Interpretation of Dreams*, Sigmund Freud primarily wanted to convince the reader that dreams had significant meaning and that interpreting them would offer insight into the wishes and desires of the unconscious mind. These insights could then be used to treat mental illnesses. Freud spends the entire first chapter of the book discussing the current state of knowledge regarding dreams. He concludes, "notwithstanding the effort of several thousand years, little progress [had] been made in the scientific understanding of dreams."[1]

Nevertheless, he begins with an overview of various theories of dreams from the past. He mentions some of these ideas only in passing—such as the ancient belief that dreams resulted from the work of spirits. He sorted the more modern ideas into several broad categories:

- Theories that related dreams to the waking state
- Theories about dreams and memory
- Theories about dreams as disturbances of sleep or as reactions to stimuli
- Theories explaining why we forget dreams so easily on awaking
- Theories about differences in function of the mind between the dreaming and waking states
- The moral content of dreams
- Theories about the functions dreams might serve
- The relationship between mental illness and dreams[2]

Devoting the balance of the text to reinforcing the importance of dreams and the insights they give us into the mind, Freud walks the reader through a sample dream and its interpretation. He notes that he learned from working with his patients "that a dream may be linked into the psychic* concatenation [sequence] which must be followed backwards into the memory from the pathological* idea as a starting point. The next step was to treat the dream as a symptom, and to apply to it the method of interpretation which had been worked out for such symptoms."[3]

> "In the following pages, I shall prove that there exists a psychological technique by which dreams may be interpreted, and that upon the application of this method every dream will show itself to be a senseful psychological structure which may be introduced into an assignable place in the psychic activity of the waking state."
>
> —— Sigmund Freud, *The Interpretation of Dreams*

Approach

The principles of the intellectual and social current known as the Enlightenment* that had been sweeping Europe since the seventeenth century emphasized the importance of reason and logic in everything. And that included the understanding and treating of psychological disorders. Most clinicians of Freud's time focused on physical factors contributing to mental disorders. As a physician, Freud had been trained in this tradition. But he was struck by the number of his patients suffering from bizarre physical symptoms that had no apparent connection to bodily dysfunction. Building on his work with Jean-Martin Charcot and Josef Breuer, both physicians who were concerned with the links between mind and behavior, Freud set out to look for alternative explanations for the types of psychological disorders he observed in some patients.

Freud became convinced that somatic—body-related—explanations remained insufficient in understanding and treating these disorders. He began to focus on the unconscious mind, and he used dream interpretation as a creative way to tap into the unconscious. For Freud, the "dream formation touches...problems of psychopathology"[*4] ("psychopathology" here means disorders of the psyche—the mind). Freud relied on both his own dreams and the dreams of his patients in developing this approach. His focus on the psychic rather than the physical, and his creative means of assessing the psychic through dream interpretation remain Freud's most unique contribution in *The Interpretation of Dreams.*

Contribution in Context

Freud was not the first to consider the unconscious mind an important factor in the development of psychological disorders. His work and correspondence with Charcot and Breuer deeply influenced him. Charcot opened Freud's eyes to the notion that some of the behavioral problems he observed could be traced back to past traumas. Indeed, "early psychoanalytic theory... is clearly indebted to Freud's encounter with Charcot."[5] From Charcot, Freud borrowed the idea that events from earlier in life can have lasting effects on the individual—even if the person has no conscious awareness of this.

Breuer treated his patients by talking to them, and this proved very effective. Freud saw this firsthand. It became an important influence as he developed his idea of conducting psychoanalysis by interpreting and analyzing dreams. Breuer believed that recreating "the memory of the incident which eventually led to [hysterical* symptoms]...[could] bring about emotional catharsis by inducing the patient to express any feeling associated with it."[6] "Catharsis" here refers to a purging or release.

Freud modified this therapeutic approach to include interpreting the dream. He found that in describing their dreams, his patients remained less likely to resist discussion of unconscious material that caused them some discomfort: "I have noticed in the course of my psychoanalytic work that the state of mind of a man in contemplation is entirely different from that of a man who is observing his psychic processes... in contemplation one exercises a critique, in consequence of which he rejects some of the ideas which he has perceived... in self-observation, on the other hand, one only has the task of suppressing the critique."[7]

---

1. Sigmund Freud, *The Interpretation of Dreams*, translated by A. A. Brill, with an introduction and notes by Daniel T. O'Hara and Gina Masucci MacKenzie (New York: Barnes & Noble Books, 2005), 13.
2. Freud, *The Interpretation of Dreams*, 13–88.
3. Freud, *The Interpretation of Dreams*, 92–93.
4. Freud, *The Interpretation of Dreams*, 3.
5. K. Libbrecht and J. Quackelbeen, "On the Early History of Male Hysteria and Psychic Trauma. Charcot's influence on Freudian Thought", *Journal of the History of the Behavioral Sciences* 31, no. 4 (1995): 370–384.
6. Richard Webster, "Freud, Charcot, and Hysteria: Lost in the Labyrinth", accessed January 3, 2016, http://www.richardwebster.net/freudandcharcot.html.
7. Freud, *The Interpretation of Dreams*, 93.

SECTION 2
IDEAS

MODULE 5
MAIN IDEAS

KEY POINTS

- The central theme in *The Interpretation of Dreams* concerns the notion that dreams contain meaningful information about the sexual and aggressive wishes of the unconscious mind.*
- Freud spends considerable time discussing how the unconscious mind disguises these wishes through the processes of condensation* (compressing several ideas into one element), displacement* (replacing one thing with another), representation* (the way the mind presents concealed thoughts in the content of the dream), and secondary elaboration* (the narrative the unconscious mind provides for what the dream contains).
- Freud's casual writing style and use of specific examples makes his arguments interesting and understandable.

Key Themes

The central idea of Sigmund Freud's *The Interpretation of Dreams* is that dreams contain important information about the desires of the unconscious mind. According to Freud, these desires are often sexual in nature. Freud cited societal views as the key reason the unconscious mind remains so preoccupied with "matters of sex" and with suppressing those impulses. Indeed, he suggests "no other impulse has had to undergo so much suppression from the time of childhood as the sex impulse in its numerous components, from no other impulse have survived so many and such intense unconscious wishes, which now act in the sleeping state in such manner as to produce dreams."[1]

Some people might experience discomfort at recognizing their unconscious sexual urges. Freud believed that portions of the unconscious mind acted as a censor to disguise this sexual material as less offensive by presenting it in dreams. Freud differentiated between a dream's manifest content* (the "story line" presented in the dream) and its latent content* (the dream's true meaning). Freud argued that it is the "latent dream content that far surpasses the manifest dream content in point of significance."[2] In other words, we can only discover the true meaning of the dream after unmasking the disguised wishes of the unconscious and we do this by analyzing and interpreting the dream. For instance, a dream about smoking a cigar might actually reflect some message about the dreamer's preoccupation with male genitalia.

> *"The more one is occupied with the solution of dreams, the more willing one must become to acknowledge that the majority of the dreams of adults treat of sexual material and give expression to erotic wishes."*
> —Sigmund Freud, *The Interpretation of Dreams*

Exploring the Ideas

Freud believed that the unconscious mind actively distorted its wishes. This distortion produced the manifest content of a dream. Specifically, Freud believed this active process of distortion involved four distinct factors: condensation, displacement, representation, and secondary elaboration.

"Condensation" reflects the notion that the mind may compress several different ideas into one single element of a dream's manifest content. As an example, Freud describes the following dream: "I have written a monograph upon a certain plant. The book lies before me, I am just turning over a folded colored plate. A dried specimen of the plant is bound with every copy as though from a herbarium [collection of plant specimens]."[3] In analyzing this dream, Freud asserts that the botanical monograph simultaneously references several things: a previous work he had written about cocaine, a friend of his who uses cocaine in his practice, a patient named Flora, the favorite flowers of his wife, his studies, and his hobbies—and all from a single element of the dream. Freud examined every element of his dream like this, producing a dizzying array of references. In this way he demonstrates that his unconscious chose the elements because "they were able to show the most extensive connections with the dream thoughts, and thus represent nuclei in which a great number of dream thoughts come together."[4]

Freud also believed that the unconscious mind engaged in "displacement". In other words, the contents of the dream might be substitutes for the latent thoughts underlying them. Using the example above, Freud indicated that one of the meanings of his dream reflected his tendency to spend too much money on his hobbies. In this case "the element 'botanical' would in no case find a place in the nucleus of dream thoughts if it were not loosely connected with it by antithesis, for botany was never among my favourite studies."[5] In dreaming of a hobby he did not enjoy, Freud believed that his unconscious was reprimanding him about spending too much money on hobbies that gave him pleasure.

When Freud speaks of "representation", he means the way in which the mind presents latent thoughts in the dream. For Freud, this usually involves some sort of visual imagery. So, for instance, the image of a man atop a tower might represent "the greatness of the man"[6] while an image of a gate "suggests a bodily opening."[7]

The final factor the unconscious mind can use to disguise the latent content of dreams involves "secondary elaboration". Here, Freud argues that the mind smooths out the dream to make its disguised messages less obvious. In other words, the dream receives a cohesive story line that belies the underlying messages and contradictions: "Those parts of the dream with which the secondary elaboration has been able to accomplish something seem to us clear; those where the power of this activity has failed seem confused."[8]

Language and Expression

Providing the reader with relevant background information, Freud offers specific examples that reinforce his argument. While readers may not agree with the conclusions he draws about the nature of the dream, they will finish the work with a greater appreciation of his thought process.

Freud wrote in German, so in talking about the language he used, we must rely on the skill of his various translators. The Austrian psychiatrist* A. A. Brill,* a contemporary of Freud's, produced the first English translation of *The Interpretation of Dreams*. Critics have noted some problems with Brill's work: some of his translations may not be completely accurate and his fluency in English fails him in places.⁹ Most of the language remains fairly straightforward, but parts of the text read more formally than modern texts; this reflects some of the idiosyncrasies of Freud's day. For example, he refers to what we commonly call a "wet dream"*—a dream that provokes ejaculation in a man—as a "pollution dream". Context helps to clarify the meaning of most of these references, but they still distract the modern reader unnecessarily. Furthermore, Brill does not translate some references to the work of other scientists and philosophers, so the reader must have some fluency in Spanish, French, and even Latin—or access to a good dictionary—to have any hope of following Freud's point.

---

1. Sigmund Freud, *The Interpretation of Dreams*, translated by A. A. Brill, with an introduction and notes by Daniel T. O'Hara and Gina Masucci MacKenzie (New York: Barnes & Noble Books, 2005), 245.
2. Freud, *The Interpretation of Dreams*, 146.
3. Freud, *The Interpretation of Dreams*, 268.
4. Freud, *The Interpretation of Dreams*, 269.
5. Freud, *The Interpretation of Dreams*, 287.
6. Freud, *The Interpretation of Dreams*, 320.
7. Freud, *The Interpretation of Dreams*, 322.
8. Freud, *The Interpretation of Dreams*, 400.
9. Freud, *The Interpretation of Dreams*, IV.

MODULE 6
SECONDARY IDEAS

KEY POINTS

* Secondary ideas in *The Interpretation of Dreams* include the notion that children experience sexual impulses towards their parents, and a rough theory of mind* (here meaning an attempt to explain the nature and functioning of the mind) outlined in chapter 7.
* These ideas foreshadow Freud's later work on the Oedipus complex* (that is, a child's desire to sleep with the parent of the opposite sex) and Freud's "stage theory" of psychosexual development*—that specific sources of sexual pleasure play a significant role in our individual development towards mental maturity.
* The rough sketch of Freud's theory of the mind provided a foundation for his later ideas about the structure of the mind and personality.*

Other Ideas

Several secondary ideas support the central thesis of Sigmund Freud's *The Interpretation of Dreams*. One is Freud's belief that as a normal part of human development, children develop a sexual attraction toward their parent of the opposite sex. This attraction creates a sense of competition with the parent of the same sex. Freud felt that many symptoms experienced by his neurotic* patients (patients exhibiting symptoms of anxiety in their specific mental condition) stemmed from this complex. He notes that "parents play a leading part in the infantile* [immature] psychology* of all later neurotics."[1] So when children play out the Oedipus complex by falling in love with one parent and hating the other, they lay the groundwork for future mental health challenges. Or, as Freud put it, they create "that fateful sum of material furnished by the psychic impulses, which has been formed during the infantile period, and which is of such great importance for the symptoms appearing in the later neurosis."*[2] But, he adds, "I do not think that neurotics are here sharply distinguished from normal human beings."[3]

In simple terms, Freud believed everyone has these feelings; neurotics simply display more exaggerated versions of them. In any case, Freud felt analyzing and interpreting dreams would offer insight into these issues.

Another important secondary idea presented in the book involves Freud's theory of the human mind, specifically how the mind operates differently in the waking and the sleeping state. He sees the mind "as a compound instrument, the component parts of which let us call...systems."[4] He describes these systems in the language of the reflex arc,* a popular concept at the time, according to which behavior can be considered a reflexive reaction to sensory input*—sensation. Freud says, "the psychic process generally takes its course from the perception end to the motility end" ("motility" refers to motion).[5] In the case of the dream, however, "the stream of thought is henceforth subjected to a series of transformations which we no longer recognize as normal psychic processes."[6]

In other words, Freud believed that we can only understand the operation of the unconscious mind,* during sleep and dreaming, by careful analysis and interpretation.

> "Perhaps we are all destined to direct our first sexual impulses towards our mothers, and our first hatred and violent wishes towards our fathers; our dreams convince us of it."
> —— Sigmund Freud, *The Interpretation of Dreams*

Exploring the Ideas

In Freud's view, we all feel sexual and aggressive impulses toward our parents. This fact remains central to some of the sexually motivated wishes of the unconscious mind. To demonstrate the eternal and universal nature of this process, Freud reaches back to ancient Greece and names this phenomenon after one of its most famous tragic figures: Oedipus.* In Freud's recounting of the myth, an oracle tells King Laius* of the city of Thebes that his yet-unborn son will murder him. When his son Oedipus is born, Laius tries to thwart this prophecy by killing the child—or rather, leaving him outside to die. The infant is rescued and taken away to grow up abroad. Later in life, Oedipus "meets King Laius and strikes him dead... and is presented with the hand of Jocasta"*—his mother.[7] Because this myth originated thousands of years earlier, Freud concluded that the Oedipus (or Oedipal) complex occurred frequently. He took this as proof that it must be a normal part of human development.

In chapter 7 of *The Interpretation of Dreams*, Freud presents his theory of the mind. He describes the mind as consisting of multiple systems, or modules. Some of these modules receive sensory information directly and control motor behavior. But

Freud's theory also involved less obvious parts of the mind. Indeed, he says that the dream itself "serves as proof for the knowledge of another part of the apparatus."⁸ Turning again to the language of reflex ("the motor end") that would be familiar to his readers, he writes that, "the last of the systems at the motor end we call the preconscious*... the system behind it we call the unconscious.*"⁹ In other words, Freud believed that a portion of the mind exists between the unconscious and the conscious. This very important idea foreshadows later writings in which he segments the mind into component parts known as the id,* ego,* and superego.* The id is the unconscious center of fundamental impulses such as sexual desire; the ego is the conscious part of the mind in which we form our social identity; the superego is the center of things such as one's sense of conscience* and morality.

In subsequent works, Freud uses the foundation laid in *The Interpretation of Dreams* to elaborate on his stage theory of psychosexual development—the notion that from infancy to maturity we pass through stages in which we derive sexual pleasure from different parts of the body, and that this process can have consequences in later life.

Overlooked

The Interpretation of Dreams brought about significant changes in the way psychiatrists* and psychologists* thought about the mind and the treatment of mental illness. As befits such an influential work, it has received an enormous amount of attention from critics and supporters over the years. So it remains difficult to imagine that any of the ideas of the text have been overlooked. But we might argue that Freud's work has been somewhat overlooked in the modern world. Indeed, more recent times have seen a "marginalization of psychoanalysis* in psychology textbooks."[10] Freud's ideas have been dismissed because he provides no empirical* evidence to support his claims (that is, no evidence verifiable by observation). This marginalization has accelerated recently as some mental health professionals have shifted their focus about the biological underpinnings of mental illness. We cannot deny that, in general, Freud has left an indelible mark on psychology and the world. But we might argue that modern-day psychologists and psychiatrists have not been greatly influenced by most of the specific ideas he presented in *The Interpretation of Dreams*.[11] However, we have seen renewed interest in some of Freud's ideas among cognitive psychologists* and neuroscientists* (scientists of the brain and nervous system). They remain interested in how we store and access certain types of memory outside our conscious awareness.

---

1. Sigmund Freud, *The Interpretation of Dreams*, translated by A. A. Brill, with an introduction and notes by Daniel T. O'Hara and Gina Masucci MacKenzie (New York: Barnes & Noble Books, 2005), 226.
2. Freud, *The Interpretation of Dreams*, 226.
3. Freud, *The Interpretation of Dreams*, 226.
4. Freud, *The Interpretation of Dreams*, 424.
5. Freud, *The Interpretation of Dreams*, 424.
6. Freud, *The Interpretation of Dreams*, 467.

7. Freud, *The Interpretation of Dreams*, 226–227.
8. Freud, *The Interpretation of Dreams*, 427.
9. Freud, *The Interpretation of Dreams*, 428.
10. Joseph Reppen, "The Relevance of Sigmund Freud for the 21st Century", *Psychoanalytic Psychology* 23, no. 2 (2006): 215–216.
11. Paul R. McHugh, "The Death of Freud and the Rebirth of Psychiatry", *The Weekly Standard*, accessed January 10, 2016, http://www.weeklystandard. com/article/12226.

MODULE 7
ACHIEVEMENT

KEY POINTS

* Freud's focus on the role of the unconscious mind* in both neurotics* and healthy individuals has profoundly influenced psychiatry* and psychology* alike.
* Freud's innovative ideas and engaging writing style appealed to a number of other clinicians and led to the establishment of psychoanalysis* as a prominent school of thought.
* The anecdotal nature of Freud's observations (their nature as "stories" applicable to certain individuals alone) and the lack of empirical*—or evidence-based—support for his theories have limited the impact of the ideas Freud introduces in *The Interpretation of Dreams*.

Assessing the Argument

In *The Interpretation of Dreams*, Sigmund Freud hoped to show that the dream serves as a product of the unconscious mind "that knows no other aim in its activity but the fulfillment of wishes."[1] He described the methods of analyzing dreams, noting that this provided a way to gain insight into the unconscious mind. Accessing unconscious thoughts would, Freud felt, give professionals an alternative means of treating various psychological disorders. It would also enhance the dreamer's self-awareness.

Freud based his method in part on the therapeutic approach that he and the Austrian physician Josef Breuer* had described in their *Studies on Hysteria* (1895).[2] In one section of the book, a patient described dreams and engaged in free association* to interpret their true meaning. Freud believed that this process helped a patient's symptoms to improve, and he reported great success when using the method. He notes that "where it has been possible to trace such a pathological* [that is, diseased] idea back to the elements of the psychic* life of the patient to which it owes its origin, this idea has crumbled away, and the patient has been relieved of it."[3] But the unscientific nature by which Freud arrived at these conclusions would provoke criticism for years to come.

Despite the critics, the ideas Freud introduced in *The Interpretation of Dreams* eventually helped establish psychoanalysis as a prominent perspective in the fields of psychiatry and psychology. In this, Freud achieved his primary aim: he drew attention to the information that dreams give us about the unconscious mind.

> "There is no better proof of the great impact Freud has made on psychiatry than the change in its scope, for which his work has been largely responsible."
> —— E. Stengel, "Freud's Impact on Psychiatry"

Achievement in Context

Freud's concept of the mind was "more profound and more precise"[4] than any of its predecessors, and as such, it appealed to many professionals in the clinical community. It took almost 10 years for the second edition of the book to be published, but over the next few decades *The Interpretation of Dreams* exploded in popularity. In fact, the publisher released six editions between 1910 and 1929 alone.[5]

Many credit Freud with establishing the psychoanalytic perspective in both psychiatry and psychology. This perspective emphasized the role of the unconscious mind in affecting conscious behavior and personality.* It became enormously influential in the early twentieth century. Indeed, one might argue that it remained one of the most prominent perspectives in the field until the mid-twentieth century.[6] One writer recalls that during his graduate studies, "Freud was prominently studied and psychoanalysts were well represented on faculties of psychology."[7]

The innovative ideas Freud proposed in such interesting ways appealed to many, who became his future disciples. His ideas about the unconscious mind and our ability to access it by interpreting dreams stood in stark contrast to the prominent ideas of the day. The Swiss thinker Carl Jung,* a student of Freud's who became an influential psychiatrist in his own right, suggests"this coinage of a theory marking itself as something extraordinary in the history of science, has a great advantage in that it stands out in bold relief as a strange and unique phenomenon against its philosophical and scientific background."[8] Freud himself recognized his achievement in changing the ways in which the clinical community viewed the unconscious. In his preface to the third edition of the book, he wrote, "the interpretation of dreams was destined to aid in the psychological analysis of the neuroses."[9] Therapists offering Freud's new talk therapy became quite prominent, especially in the

United States.

Limitations

Despite its successes, the limitations of *The Interpretation of Dreams* ultimately curtailed its reach and diminished its influence. The modest initial sales of the book disturbed Freud, as they would any author. In his preface to the book's second edition, he writes, "If there has arisen a demand for a second edition...I owe no gratitude to the interest of the professional circles to whom I appealed."[10] However, the increasing frequency of later editions of the book pointed to a surge in interest.

The most fundamental limitations of *The Interpretation of Dreams* concern the ways in which Freud developed his ideas about the unconscious mind. He relied on self-analysis and anecdotal reports in formulating the theory, rather than on systematic, empirical (evidence-based) observations. The ideals of seventeenth-and eighteenth-century intellectual movement known as the Enlightenment* stressed the importance of reason, logic, and scientific inquiry. While his contemporaries upheld these ideals, Freud's anecdotal, unverifiable theory seemed to abandon them. This called into question the validity of Freud's work. Indeed, some viewed Freud as "either an evil-minded villain or someone with a diseased brain pretending that his own delusions* were clinical observations."[11]

1. Sigmund Freud, *The Interpretation of Dreams*, translated by A. A. Brill, with an introduction and notes by Daniel T. O'Hara and Gina Masucci MacKenzie (New York: Barnes & Noble Books, 2005), 446.
2. Sigmund Freud and Josef Breuer, *Studies on Hysteria*, trans. James Strachey (London: Hogarth Press, 1955).
3. Freud, *The Interpretation of Dreams*, 92.
4. Ernest Jones, "Freud and His Achievements", *The British Medical Journal* 1, no. 4974 (1956): 997–1000.
5. "Freud's book, 'The Interpretation of Dreams' released 1900", *PBS*, accessed January 24, 2016, http://www.pbs.org/wgbh/aso/databank/ entries/dh00fr.html.
6. "Introduction to Psychology", in *Psychology*, 1–34 (Houston, TX: OpenStax College, 2014).
7. Joseph Reppen, "The Relevance of Sigmund Freud for the 21st Century", *Psychoanalytic Psychology* 23, no. 2 (2006): 215–216.
8. C. G. Jung, "Sigmund Freud in His Historical Setting", *Journal of Personality* 1, no. 1 (1932), 48–55.
9. Freud, *The Interpretation of Dreams*, 7.
10. Freud, *The Interpretation of Dreams*, 5.
11. Jones, "Freud and His Achievements", 998.

MODULE 8
PLACE IN THE AUTHOR'S WORK

KEY POINTS
- Freud spent his career describing the importance of the unconscious mind* in behavior and personality,* and he founded the psychoanalytic* school of thought in psychology* and psychiatry.*
- *The Interpretation of Dreams* laid the foundation for Freud's later theories.
- Many consider *The Interpretation of Dreams* one of Freud's most significant achievements. It established his reputation as the father of psychoanalysis.

Positioning

Before Sigmund Freud wrote *The Interpretation of Dreams*, his best-known work had been the 1895 book *Studies on Hysteria*, which he coauthored with the Austrian physician Josef Breuer.*[1] Critics hailed it as a "landmark in psychopathology"* and consider that its publication date marks "the inception of psycho-analysis."[2] In it, Freud and Breuer describe patient case studies and document a unique therapeutic approach. Their new form of "talking therapy"—allowing patients to discuss their issues—was found to relieve a patient's neurotic* symptoms. These outcomes so intrigued Freud that he felt the need to "press forward on the path taken by Breuer until the subject has been fully understood."[3]

Five years later, Freud published *The Interpretation of Dreams*. It reflected ideas he had developed during and after his time with Breuer. Freud had always been fascinated with the role the unconscious mind plays in behavior and personality development. The book's focus on using dreams to access the wishes of the unconscious mind arose from this interest. *The Interpretation of Dreams* laid a foundation for much of the rest of his professional work. Several ideas it introduced foreshadowed Freud's later writings about both his theory of mind* (his theory of the nature, structure, and functioning of the mind) and his "stage theory" of psychosexual development,* according to which we achieve psychic* maturity from infancy by passing through various stages in which we derive sexual pleasure from different bodily sources.

> "Freud's life-work may be broadly summarized as the exploration of the unconscious after he had devised a method, now known as psycho-analysis for doing so."
> —— Ernest Jones, "Freud and His Achievements"

Integration

Throughout his career as a psychoanalyst, Freud believed that to understand human behavior and personality, one must understand the wishes of the unconscious mind. All Freud's work centers on this concept, using *The Interpretation of Dreams* as a foundation. In *The Interpretation of Dreams*, Freud presented a rough, reflex-like sketch of the mind. In later works, he outlined "a 'topography' of the psyche* in three parts: the id,* the ego,* and the super-ego.*"[4] These he identified as divisions of the unconscious mind: together, they contain all the desires and wishes an individual's mind has suppressed. Freud believed that most of those desires remained sexual in nature. The ego essentially represents the conscious mind.* It is "sandwiched between the id and super-ego."⁵ The superego represents the sense of conscience.* The id represents the sexual and aggressive impulses of the unconscious mind. According to Freud, the ego remains in a constant state of tension trying to balance the demands of both the id and the superego.

In *The Interpretation of Dreams*, Freud hints at another major theory to come. In describing the sexual and aggressive impulses that children feel toward their parents, Freud references the story of the mythical Greek King Oedipus.* This story found a place in his later works, when Freud proposes a stage theory of psychosexual development. One of these stages includes the phenomenon Freud named the "Oedipus complex".* He sees this as a normal part of every person's development. Briefly, the stage theory of psychosexual development describes development migrating through erogenous zones* (places in which sexual pleasure is to be gained) in a growing child's body.

Freud associates each zone with a series of conflicts; people who fail to successfully resolve these conflicts develop fixations* (excessive preoccupations) that affect their personality and behavior. For example, one of the earliest stages involves the erogenous zone centering on the mouth. This would correlate to early in an infant's life, when it relies on suckling to survive. If the mother weaned the child too soon or too late, an oral fixation could develop. Accordingly, "an adult who smokes, drinks, overeats, or bites [their] nails"[6] would be assumed to suffer from an oral fixation. As time goes on, the erogenous zone moves from the mouth to the anus and then on to the genitals or phallus (the period when the Oedipus complex occurs). After the phallic stage, a period of relative inactivity ensues. Then, about the time of puberty, the erogenous zone becomes centered on the genitals and remains there for the rest of the individual's life.

Significance

A prolific writer, Sigmund Freud published more than 300 works over the course of his career. Yet he often identified *The Interpretation of Dreams* as his personal favorite.[7] Indeed, "the preface to the third English edition (1931) of *The Interpretation of Dreams* makes clear the place of this book in the eye of its author: '[This book] contains...the most valuable of all the discoveries it has been my good fortune to make. Insight such as this falls to one's lot but once in a lifetime.'"[8] Both supporters and critics share this view, routinely citing "the centrality of the dream book to psychoanalysis."[9] Supporters have levied "laudatory epithets [that is, admiring descriptions], with [the German American historian] Peter Gay* comparing it to [the foundational evolutionary theorist] Charles Darwin's* *On the Origin of Species* as a 'revolutionary classic shaping modern culture.'"[10]

Earlier in his career, Freud published on topics ranging from a historical and clinical study of cocaine to a study of cerebral paralysis (paralysis of the brain) in children.[11] After *The*

Interpretation of Dreams, his work followed one direction as he continued to elaborate on many of the ideas he introduced in his masterwork. *The Interpretation of Dreams* represented Freud's "magnum opus [masterpiece]... which was the foundation of all his later work."[12] While Freud's influence has waned in the last few decades, that does not diminish the central role this text played in his career.

---

1. Sigmund Freud and Josef Breuer, *Studies on Hysteria*, trans. James Strachey (London: Hogarth Press, 1955).
2. Ernest Jones, "Freud and His Achievements", *The British Medical Journal* 1, no. 4974 (1956): 997–1000.
3. Sigmund Freud, *The Interpretation of Dreams*, translated by A. A. Brill, with an introduction and notes by Daniel T. O'Hara and Gina Masucci MacKenzie (New York: Barnes & Noble Books, 2005), 92.
4. Paul Ricoeur, "Sigmund Freud", in Karl Simms, *Paul Ricoeur: Routledge Critical Thinkers* (Abingdon, UK: Taylor & Francis, 2002), 46.
5. Ricoeur, "Sigmund Freud", 46.
6. "Personality", in *Psychology*, 369–410 (Houston, TX: OpenStax College, 2014), 375.
7. Kendra Cherry, "Books by Sigmund Freud: Freud's Most Famous and Influential Books", accessed January 9, 2016, http://psychology.about.com/ od/sigmundfreud/tp/books-by-sigmund-freud.html.
8. Patricia Kitcher, *Freud's Dream: A Complete Interdisciplinary Science of Mind* (Cambridge, MA: MIT Press, 1992), 113.
9. Kitcher, *Freud's Dream*, 113.
10. Kitcher, *Freud's Dream*, 113.
11. Jones, "Freud and His Achievements", 997–1000.
12. Jones, "Freud and His Achievements", 999.

SECTION 3
IMPACT

MODULE 9
THE FIRST RESPONSES

KEY POINTS
* Freud's emphasis on infantile* sexuality and his nonscientific approach served as fodder for early critics.
* Freud ignored some criticism, but when former colleagues started criticizing his work, he often cut off contact with them.
* How readers received the text largely depended on their personal convictions and their loyalty to Freud.

Criticism

Critics largely ignored Sigmund Freud's *The Interpretation of Dreams* in the first few years after its publication because it did not accord with the common scientific themes of the time but its popularity grew. As Freud's ideas gained momentum, he found himself at the head of a new school of thought in the psychiatric* and psychological* communities: the psychoanalytic* movement. The First International Congress of Psychoanalysis convened in 1908 and a group of psychoanalysts established the International Psychoanalytical Association two years later. Freud became something of a celebrity. He and his student, the Swiss psychoanalyst Carl Jung,* traveled to the United States, lecturing on psychoanalysis.¹

Despite all this acclaim, the book also generated controversy and criticism. Freud's critics took issue with the speculative nature of his ideas and the subjectivity involved in developing them. Seeing these as major flaws, critics questioned the scientific value of Freud's work. Some merely distanced themselves from Freud; others openly criticized his ideas.

Josef Breuer,* Freud's collaborator on *Studies on Hysteria* (1895), both distanced himself and criticized Freud. Indeed, scholars have traditionally assumed that he severed all ties with Freud "because Breuer objected to Freud's claim that there was a sexual etiology* for the psychoneuroses."*[2] ("Etiology" refers to the causes of a disease.) Others have claimed that the two broke off relations because "Breuer...could not tolerate such an affront to rationalism"* inherent in Freud's ideas.[3]

Breuer was not the only person in Freud's life who switched from colleague to critic. Carl Jung, one of Freud's students, also turned into a fierce critic. "The dynamic that developed between them was one where, having first declared Jung his 'heir apparent', Freud would go on to call him 'mad' six years later."[4] This separation occurred when Jung took exception to Freud's ideas about the importance of sexual impulses and early childhood experiences in determining adult behavior. Jung countered that human goals and aspirations also serve as important motivating forces. Jung considered Freud's ideas about the unconscious mind* underdeveloped. While his view of the personal unconscious was very similar to Freud's, Jung identified an extra layer of the unconscious mind, which he called the "collective unconscious"*[5]—a part of the unconscious mind shared or contributed to by everyone. Another factor differentiating Jung from his former mentor is his treatment of religion and spirituality. Jung placed these in the collective unconscious.[6]

> *"The Freudian theory...is at best a partial truth, and therefore in order to maintain itself and be effective, it has the rigidity of a dogma and the fanaticism of an inquisitor."*
> ——Carl Jung, "Sigmund Freud in His Historical Setting"

Responses

Describing Freud's response to his critics, one writer notes that he "bore all this hostility with considerable fortitude and never deigned to reply to it in public; in private, he even derived a certain amount of amusement... I remember his remarking once:'My opponents may abuse my doctrines by day, but I am sure they dream of them by night.'"[7] This aloof approach did little to abate the criticism. Indeed, to this day critics continue to raise many of the same objections about the unscientific nature of Freud's theories.

Freud could use humor to distance himself from some opponents, but criticism from colleagues who had once been his closest allies wounded him deeply. Freud felt personally betrayed and hurt by Jung's criticism. What had begun as a great friendship steeped in mutual respect transformed into a fierce rivalry. Jung and Freud's correspondence became increasingly hostile; eventually they stopped communicating altogether. Indeed, in his last letter to Jung, Freud wrote, "I propose that we abandon our personal relationship entirely. I shall lose nothing by it, for my only emotional tie with you has long been a thin thread."[8] As a consequence, Freud began isolate himself from his critics. Instead, he surrounded himself with a close-knit group of colleagues who fiercely defended the principles of psychoanalysis.⁹

Conflict and Consensus

Freud and his critics could not agree on much, and these disagreements continued for many years. Jung and Breuer were not the only former colleagues who felt Freud's professional and personal wrath. He also broke with the Austrian physician and psychotherapist Alfred Adler* and the Hungarian psychoanalyst Sándor Ferenczi,* both of whom were former students of his. Ernest Jones,* a psychoanalyst who supported Freud to the end, says Adler and Ferenczi became unable "to face the conclusions that follow from deep psycho-analytic investigation... [They] turned away and repudiated the conclusions they had once accepted and expounded."¹⁰ However, Jones admits that like Adler and Jung, many critics "were able to enjoy successful careers on the basis of rejecting Freud's sexual theories, and the general public welcomed them with a sense of relief at being rescued from Freud's displeasing ideas."¹¹

In the end, reactions to Freud's *The Interpretation of Dreams* depended largely on personal convictions and loyalties. Some, like Jones, remained incredibly loyal; in fact, Jones went on to become Freud's biographer. Theorists like Jung accepted many of the doctrines Freud set out, but modified them in ways that matched their own ideas about the nature of the unconscious mind. Breuer and others dismissed Freud's ideas outright. They objected to the unscientific way in which he constructed his theories of the unconscious mind and dream analysis.*

1. Sigmund Freud, *The Interpretation of Dreams*, translated by A. A. Brill, with an introduction and notes by Daniel T. O'Hara and Gina Masucci MacKenzie (New York: Barnes & Noble Books, 2005), xv.
2. John P. Muller, "A Re-Reading of *Studies on Hysteria*: The Freud-Breuer Break Revisited", *Psychoanalytic Psychology* 9, no. 2 (1992): 129–156.
3. Muller, "A Re-Reading of *Studies on Hysteria*", 129.
4. Hester McFarland Solomon, "Freud and Jung: An Incomplete Encounter", *Journal of Analytical Psychology* 48, no. 5 (2003): 553–569.
5. Kendra Cherry, "Sigmund Freud Photobiography: Freud and Jung", accessed January 9, 2016, http://psychology.about.com/od/sigmundfreud/ig/Sigmund-Freud-Photobiography/Freud-and-Jung.htm.
6. "The Well-Documented Friendship of Carl Jung and Sigmund Freud", Historacle.org, accessed January 9, 2016, http://historacle.org/freud_jung.html.
7. Ernest Jones, "Freud and His Achievements", *The British Medical Journal* 1, no. 4974 (1956): 997–1000.
8. David Eidenberg, "Freud and Jung: A 'Psychoanalysis' in Letters", *Psychological Perspectives* 57 (2014): 7–24.
9. Cherry, "Sigmund Freud Photobiography: Freud and Jung".
10. Jones, "Freud and His Achievements", 998.
11. Jones, "Freud and His Achievements", 998.

MODULE 10
THE EVOLVING DEBATE

KEY POINTS

* *The Interpretation of Dreams* helped establish psychoanalysis* as both a therapeutic technique and a school of thought.
* Psychodynamic* approaches (those approaches linking the unconscious mind* and conscious behavior) like the analytical perspective* owe their establishment to the works of Freud.
* Scholars consider *The Interpretation of Dreams* Freud's greatest work; it remains core reading for anyone trying to better understand the basic principles of psychoanalysis.

Uses and Problems

Many critics dismissed Sigmund Freud's *The Interpretation of Dreams*. They felt he "had exaggerated the role of sexuality, relied on faulty methods, and succumbed to wild speculation."[1] However, his supporters recognized—and continue to recognize—the enormous impact his work has had on both psychiatry* and psychology.*

The Interpretation of Dreams laid the foundation for the psychoanalytic perspective and established Freud as the unquestioned leader of the psychoanalytic movement. Freud's perspective created a new way of thinking—"a cultural revolution comparable in scope to that unleashed by [the English naturist and evolutionary theorist] Charles Darwin.*"[2] It also offered a new therapeutic technique for professionals. Soon, psychiatrists began to treat people with various mental illnesses by using therapies involving dream analysis* and free association.*[3]

Freud remained fiercely protective of the key tenets of his theories. He especially insisted on the role the sexual impulse plays in fueling the desires of the unconscious mind. This idea remained central to him. Freud's rigid adherence to his ideas created tensions with some of his strongest supporters, a number of whom wanted to modify the concept of the unconscious. Freud's former student Carl Jung* and others eventually founded their own related schools of thought. Derived from Freudian principles, they became collectively known as neo-Freudians* and developed distinguished careers in their own right.

> "Freud has not only expanded the range of psychiatry, he has also given it new dimensions in depth by the discovery and exploration of the dynamic unconscious—that is, of mental forces which influence behaviour and which can be brought into consciousness.*"
>
> —— E. Stengel, "Freud's Impact on Psychiatry"

Schools of Thought

The psychoanalytical perspective born of Freud's work emphasized the ways in which the unconscious mind affects behavior and personality.* Freud viewed the unconscious mind as a reservoir of unfulfilled wishes and impulses. People express their unconscious sexual wishes in dreams. In fact, in his view, most of the psychic* energy tied up in the unconscious mind directly results from sexual impulses being forced out of conscious awareness by some censor mechanism. Freud would later name this internal censor the "superego"*—the part of the mind governing morality and our unconscious instincts. Freud held that we can only truly understand these unconscious wishes by engaging in the processes of dream analysis and interpretation.[4]

As happens with any innovator, not all of Freud's followers accepted all of his ideas. Some of these disagreements led to the development of alternate schools of thought. Scholars credit Freud's student-turned-critic Carl Jung with establishing the analytical perspective—an alternative school of thought to psychoanalysis. Jung emphasized his own unique ideas about the collective unconscious*—the part of the unconscious mind we share with others—and its shared images and patterns, which he called "archetypes".* Unlike Freud, Jung also applied many of his ideas to religion and spirituality.[5] Jung might have remained a supporter of Freud had his teacher been open to modifying his approach. However, "in expecting that Jung, as his adept and chosen heir, would display uncritical devotion to the theory of psychoanalysis that he had conceived, Freud had badly misjudged the younger psychiatrist."[6]

In Current Scholarship

Today's scholars often group the psychoanalytic and analytical perspectives together with others under the heading of "psychodynamic" approaches. The psychodynamic perspective refers to "all the theories in psychology that see human functioning based upon the interaction of drives and forces within the person, particularly unconscious, and between the different structures of the personality."[7] While the influence of psychoanalysis in particular has waned in recent years, many mental health professionals still follow the therapeutic approaches put forward by Freud and his followers, the neo-Freudians. Anyone who wants to understand this perspective would be wise to read this critical text. One clinician noted that she "cannot be a fully 'Jungian analyst' without having an intimate knowledge of and without pursuing in depth my study of the foundations and theoretical developments of 'Freudian psychoanalysis.'"[8]

Outside clinical circles, people still discuss *The Interpretation of Dreams* and Freud's other works widely. These discussions generally focus on historical value and influence. Still, new advances in cognitive neuroscience*—scientific investigation into the process of thought—may open the door to a more prominent role for Freud's works today. One author notes that "within the context of certain psychiatric and neuroscientific* circles, psychoanalysis has become a topic of renewed interest."[9] Indeed, scholars resumed defending "the potential legitimacy of psychoanalytic theory and practice" as early as the mid-1980s.[10]

---

1. Anthony D. Kaunders, "Truth, Truthfulness, and Psychoanalysis: The Reception of Freud in Wilhelmine Germany", *German History* 31, no. 1 (2013): 1–22.
2. Henri F. Ellenberger, *The Discovery of the Unconscious: The History and Evolution of Dynamic Psychiatry* (New York: Basic Books, 2008), 418.
3. Ernest Jones, "Freud and His Achievements", *The British Medical Journal* 1, no. 4974 (1956):

997–1000.
4. Saul McLeod, "Psychodynamic Approach", *Simply Psychology*, accessed January 10, 2016, http://www.simplypsychology.org/psychodynamic.html.
5. McLeod, "Psychodynamic Approach".
6. Hester McFarland Solomon, "Freud and Jung: An Incomplete Encounter", *Journal of Analytical Psychology* 48, no. 5 (2003): 554.
7. McLeod, "Psychodynamic Approach".
8. Solomon, "Freud and Jung," 553–569.
9. Nima Bassiri, "Freud and the Matter of the Brain: On the Rearrangements of Neuropsychoanalysis", *Critical Inquiry* 40, no. 1 (2013): 83–108.
10. Bassiri, "Freud and the Matter of the Brain", 83.

MODULE 11
IMPACT AND INFLUENCE TODAY

KEY POINTS

* While *The Interpretation of Dreams* has become less relevant for today's mental health professionals, it remains a classic text still capable of capturing the interests of modern-day psychologists.*
* Freud's insistence on the importance of the unconscious* brought into focus the unconscious aspects of the mind and has left an indelible mark on the history of psychology.
* Lively debate continues about the relevance of Freud's ideas in the twenty-first century. Some relegate his ideas to the annals of history and others point to the potential value that these ideas still hold.

Position

Sigmund Freud's *The Interpretation of Dreams* ushered in the psychoanalytic* movement. "[From] 1935 to 1975... Freudianism was the unchallenged doctrine of American psychiatry.*"[1] One writer lamented that "recent discoveries in biomedicine,* which the public may think are great advances, have in fact plucked the 'soul' from psychiatry,* leaving it a cold business that dispenses magical pills rather than addressing patients in all their tragic peculiarity."[2] ("Biomedical" approaches to mental health are based on the treatment of the physical organism, routinely conducted with medicine.) Others suggest that this change allows psychiatry to "grow as a science-based, evidence-driven discipline."[3]

Today, mental health professionals use a range of therapeutic approaches. Psychoanalysts still offer their services, but their numbers—and their influence—have steadily declined. Indeed, it has been claimed that "the number of publishers willing to publish psychoanalytic titles has decreased dramatically... the major university presses are no longer willing even to consider psychoanalytic titles."[4]

Some have attributed this shift in attitude to "the biological turn in psychiatry and emergence of cognitive neurosciences* as the dominant paradigm in psychology [departments] and journals around the world."[5] Cognitive neurosciences examine thought in the light of the electrical and chemical functioning of the physical brain. Although this once seemed a death knell for psychoanalysis, it may become an opportunity for its rebirth. As the interest in neuroscience "has grown steadily in the last decades, configuring a well-developed field of knowledge... some of its practitioners have paid increasing attention to areas of inquiry within, or close to, the realm of psychoanalytical interest."[6] Psychoanalysis and neuroscience could potentially work together, with possible opportunities for joint investigation involving such concepts as memories that exist outside conscious awareness.

> "In their disciplinary merger, it has been argued, psychoanalysis and the neurosciences could directly benefit one another; the neurosciences could be infused with a more robust theory of subjective experience while psychoanalysis could make the transition to becoming a testable, experimental science."
> ——Nima Bassiri, "Freud and the Matter of the Brain: On the Rearrangements of Neuropsychoanalysis"

Interaction

Not everyone feels enthusiastic about the idea of merging psychoanalysis and neuroscience. In fact, many in the psychoanalytic camp have resisted it, considering psychoanalysis "a 'nonscientific field' of knowledge, linked to humanities* and far from natural sciences.*"[7] The "humanities" is a broad area of scholarship including the studies of history, literature, and culture, and so by linking psychoanalysis with neuroscience, some feel that its very foundation would become threatened.

However, the field of psychoanalysis remains quite splintered. Indeed, it has been argued that, "a multitude of groups and factions...debate the true heritage of the Freudian legacy, and unfortunately dedicate insufficient time and effort to improving teaching and research methods."[8] As a result, "psychoanalysts have disappeared from the academic frontline in many countries, especially in the United States, where they once occupied privileged positions in many departments."[9]

A number of psychoanalysts have even begun to question the relevance of Freud's ideas in the twenty-first century. One writer notes, "there are 'two Freuds'... Freud, the natural scientist,* and Freud the hermeneutist."[10] A hermeneutist is someone who specializes in interpretation—in other words, we may think of Freud as both a physician and an expert in interpretation, and specifically in interpreting the individual's unfulfilled wishes as expressed in dreams. The writer quoted above argues that to determine the relevance of Freud's arguments, we must examine them on a case-by-case basis.

The Continuing Debate

Neuroscientists may begin to examine memories stored in the unconscious mind. Many psychoanalysts welcome this possibility. Applying psychoanalytic principles in the context of neuroscience allows us to "study the physical support for mental life, which helps us complement and refine our ideas, and in addition allows us to offer models and concepts that may enrich the research of biologists."[11] In other words, today's technology can show us how unconscious processes produce activity in the brain.

However, neuroscientists will investigate these ideas in very different ways than their psychoanalyst colleagues. As one writer has asserted, "Freud's biologically based constructs...are not relevant for the twenty-first century... [As] it turns out psychoanalysts have begun to view the effects of biology on psychological reactions, but a biology based on neuroscience and the structures of the brain, not the neurophysiology* of the functioning of the nerves and inherited content of the mind, that guided Freud's thinking."[12] "Neurophysiology" here refers to the physical structure and functioning of the brain and nervous system.

Still, neuroscience may offer an unanticipated lifeline to psychoanalysis. Unless it establishes its relevance for today, the discipline will fall victim to "its own inevitable obsolescence, a consequence of its inability or refusal to evolve as a natural, experimental science."[13] We may note some irony here: in establishing psychoanalysis, Freud, who trained as a medical doctor, decided to forego scientific methods of testing and verification. He relied instead on anecdotes and personal observation. However, if psychoanalysis is to remain relevant, it must now rely on science.

---

1. Paul R. McHugh, "The Death of Freud and the Rebirth of Psychiatry", *The Weekly Standard*, accessed January 10, 2016, http://www.weeklystandard.com/article/12226.
2. McHugh, "The Death of Freud and the Rebirth of Psychiatry".
3. McHugh, "The Death of Freud and the Rebirth of Psychiatry".
4. Carlo Strenger, "Can Psychoanalysis Reclaim the Public Sphere?", *Psychoanalytic Psychology* 32, no. 2 (2015): 293–306.
5. Strenger, "Can Psychoanalysis Reclaim the Public Sphere?", 294.
6. Miguel Angel Gonzalez-Torres, "Psychoanalysis and Neuroscience. Friends or Enemies?", *International Forum of Psychoanalysis* 22, no. 1 (2013): 35–42.
7. Gonzalez-Torres, "Psychoanalysis and Neuroscience", 36.
8. Gonzalez-Torres, "Psychoanalysis and Neuroscience", 37.
9. Gonzalez-Torres, "Psychoanalysis and Neuroscience", 37.
10. George Frank, "A Response to 'The Relevance of Sigmund Freud for the 21st Century'", *Psychoanalytic Psychology* 25, no. 2 (2008): 375–379.
11. Gonzalez-Torres, "Psychoanalysis and Neuroscience", 38.
12. Frank, "A Response", 376.
13. Nima Bassiri, "Freud and the Matter of the Brain: On the Rearrangements of Neuropsychoanalysis", *Critical Inquiry* 40, no. 1 (2013): 83–108.

MODULE 12
WHERE NEXT?

KEY POINTS
* The best future potential for *The Interpretation of Dreams* lies in the ways it can support neuropsychoanalytic* interests (neuropsychoanalysis is the term used to describe the merging of psychoanalytic* and neuroscience* perspectives; neuroscience is the scientific study of the brain and nervous system).
* *The Interpretation of Dreams* will continue to serve as an important part of the history of both psychology* and psychiatry.*
* *The Interpretation of Dreams* remains one of the core—and most influential—texts in both psychology and psychiatry.

Potential

Sigmund Freud's *The Interpretation of Dreams* has shaped unique schools of thought in the fields of psychology and psychiatry. In this sense, it has already achieved its full potential. Psychoanalysis remained a dominant perspective in psychology for several decades, its influence perhaps most apparent in the United States. Since the mid-1970s, the numbers—and influence—of psychoanalysts have declined significantly. As pharmacological treatments of various psychological disorders have become available, interest has shifted to more evidencebased forms of therapy (pharmacology is the science of drugs).[1] It remains difficult to imagine any future situation in which psychoanalysis might regain its tremendous influence. But *The Interpretation of Dreams* will certainly retain its place as one of the classic texts in both fields. Practitioners and lay readers will continue to study it for its significant historical value.

That said, neuroscientists trying to better understand the mind have renewed their interest in some of Freud's psychoanalytic ideas. This has led to an increased number of calls to merge the two disciplines into the new field of neuropsychoanalysis.* This discipline has attracted some significant support. The Nobel Prize-winning neuropsychiatrist* Eric Kandel* said, "it is my hope that by joining with cognitive neuroscience* in developing a new and compelling perspective on the mind and its disorders, psychoanalysis will regain its intellectual energy."[2] Such a merger may represent the best future potential for Freud's ideas.

> *"My purpose... is to suggest one way that psychoanalysis might re-energize itself, and that is by developing a closer relationship with biology in general and with cognitive neuroscience in particular."*
> —— Eric R. Kandel, "Biology and the Future of Psychoanalysis: A New Intellectual Framework for Psychiatry Revisited"

Future Directions

In a 1999 paper, Kandel laid out a framework for the future of neuropsychoanalysis. "Central to psychoanalysis," Kandel writes, "is the idea that we are unaware of much of our mental life."[3] Neuroscientists already make distinctions between explicit (conscious) memories* and implicit or procedural memories.*

"Procedural" here relates to unconscious memories such as those concerning the performance of processes such as tasks.

Neuropsychoanalysts might map these distinctions onto the psychoanalytic view of the mind. Kandel points out that "one of the earlier limitations to the study of unconscious psychic* processes was that no method existed for directly observing them." But "a key contribution that biology can now make—with its ability to image mental processes and its ability to study patients with lesions in different components of procedural memory—is to change the basis of the study of the unconscious mental processes from indirect inference to direct observation."[4]

Kandel notes that some of the techniques Freud used, such as free association,* also lend themselves to biological study. "[We] have in biology a good beginning of an understanding of how associations develop in procedural memory...insofar as aspects of procedural knowledge are relevant to moments of meaning, these biological insights should prove useful for understanding the procedural unconscious."[5] He sees further potential convergences between the psychoanalytic and neuroscience perspectives.These include the links between thoughts and psychopathology,* the role that early experiences play in psychopathology, and the role played by the part of the brain known as the prefrontal cortex,* associated with higher cognitive functioning, in preconscious* thought. "Preconscious" here refers to the part of the mind that exists between the unconscious and the conscious; memories the mind does not suppress exist here before they are called into consciousness.*

Summary

Sigmund Freud's *The Interpretation of Dreams* remains a classic text in the fields of both psychology and psychiatry. We cannot overstate the impact that Freud, this important book, and the subsequent rise of the psychoanalytic school of thought have had in these areas. Kandel sums it up well when he writes, "During the first half of the twentieth century, psychoanalysis revolutionized our understanding of mental life. It provided a remarkable set of new insights about unconscious mental processes."[6] Scholars have compared Freud's genius to that of the English evolutionary scholar Charles Darwin* and the German physicist Albert Einstein.* Although Freud wrote prolifically, scholars have identified *The Interpretation of Dreams* as his single most important work.[7] Any reader interested in the history of psychology and psychiatry, in neuroscience, or in the ways Freud has influenced the culture at large would benefit from reading *The Interpretation of Dreams*. And as neuroscientists renew their interest in psychoanalytic principles, this text should remain relevant for years to come.

There will always be new insights to gain—indeed, many concepts of the unconscious mind* have changed in the last few decades. But *The Interpretation of Dreams* sets forth one of the most famous and influential theories of mind* ever proposed. While its relevance may have changed since Freud's day, its significance is assured.

1. Paul R. McHugh, "The Death of Freud and the Rebirth of Psychiatry", *The Weekly Standard*, accessed January 10, 2016, http://www.weeklystandard.com/article/12226.
2. Eric R. Kandel, "Biology and the Future of Psychoanalysis: A New Intellectual Framework for Psychiatry Revisited", *The American Journal of Psychiatry* 156, no. 4 (1999): 505–524.
3. Kandel, "Biology and the Future of Psychoanalysis", 508.
4. Kandel, "Biology and the Future of Psychoanalysis", 510.
5. Kandel, "Biology and the Future of Psychoanalysis", 510.
6. Kandel, "Biology and the Future of Psychoanalysis", 505.
7. Patricia Kitcher, *Freud's Dream: A Complete Interdisciplinary Science of Mind* (Cambridge, MA: MIT Press, 1992).

GLOSSARY OF TERMS

1. **Analytical perspective:** an alternative school of thought to psychoanalysis founded by Freud's student Carl Jung.
2. **Anti-Semitism:** prejudice targeting Jewish people.
3. **Archetype:** in this context, refers to the images and patterns that occur in the collective unconscious (a term coined by the psychoanalyst Carl Jung for the part of the unconscious mind shared among populations or communities).
4. **Biomedicine:** the application of biological principles in medicine.
5. **Collective unconscious:** Jung's concept of the part of the unconscious mind all people share.
6. **Condensation:** in this context, the compression of several ideas into one element of the manifest (obvious) dream content.
7. **Cognitive neuroscience:** a branch of neuroscience that studies behavior and thought processes in terms of the biological functions of the brain.
8. **Cognitive psychology:** the study of the human mind and behavior as it relates to thought.
9. **Conscience:** one's sense of right and wrong.
10. **Consciousness:** one's state of awareness.
11. **Defenses (ego defense mechanisms):** unconscious techniques for dealing with anxiety.
12. **Delusion:** a false belief.
13. **Displacement:** in this context, substitutions made in the latent (concealed and "true") and manifest (describable) content of dreams.
14. **Dream analysis:** in this context, a process that uses free association to determine the meaning of a dream.
15. **Ego:** according to Freud, the part of the conscious mind that operates between the id and the superego.
16. **Enlightenment:** a seventeenth-and eighteenth-century cultural, intellectual, and philosophical movement that emphasized social and personal progress

through education, science, individualism, and reason.
17. **Empirical:** based on observable evidence.
18. **Erogenous zone:** according to Freud, the area of the body in which sexual energy is centered. This zone moves through various areas as a person develops.
19. **Explicit memory:** memories a person is consciously aware of and can recall.
20. **Etiology:** the reason something occurs.
21. **Fixation:** persistent focus on an erogenous zone that was unsuccessfully resolved during psychosexual development.
22. **Free association:** an exercise in which one person (the therapist, for instance) says a word and the other (the patient) responds immediately with another word. There may be no obvious connection between the two words.
23. **Freudian slip:** a mistake in speech that is thought to reveal something about the individual's unconscious wishes.
24. **Humanities:** disciplines in academia that focus on human culture.
25. **Hypnosis:** a procedure in which a person experiences an altered state of consciousness that renders him or her especially open to the power of suggestion.
26. **Hysteria:** in this context, a neurotic disorder characterized by extreme emotion and altered sensory/motor function.
27. **Id:** a reservoir of unconscious impulses and psychic energy.
28. **Inertia:** in this context, a discharge of excess excitation in the nervous system.
29. **Infantile:** relating to infants.
30. **Latent content:** the true meaning of a dream.
31. **Manifest content:** a dream's story line.
32. **Materialism:** in this context, the assumption that all mental processes are a function of the brain.

33. **Motor output:** signals sent to muscles from the nervous system that cause the body to move or act.
34. **Natural science:** academic disciplines concerned with understanding the natural world.
35. **Nazi Germany (1933—1945):** the period when Adolf Hitler and his Nazi Party ruled Germany and, during the course of World War II, several other countries as well. Fiercely anti-Semitic, the Nazis would eventually imprison and execute six million Jews, homosexuals, and others they considered "undesirable".
36. **Neo-Freudians:** a label used to describe theorists and therapists influenced by Freud but who took exception to some of his specific assertions about psychoanalysis.
37. **Neural activity:** activity in the cells of the brain.
38. **Neurologist:** a doctor specializing in the treatment of diseases of the nervous system and brain.
39. **Neuropathology:** the branch of medicine that deals with diseases of the nervous system.
40. **Neurophysiology:** study of the function of the nervous system.
41. **Neuropsychiatry:** the study of the role of the nervous system in disorders of the mind.
42. **Neuropsychoanalysis:** the term used to describe a merging of psychoanalytic and neuroscience perspectives.
43. **Neuroscience:** an interdisciplinary area concerned with studying the structure and function of the nervous system.
44. **Neurosis (psychoneurosis):** mental illness involving anxiety.
45. **Neurotics:** patients exhibiting symptoms of anxiety in their specific mental condition.
46. **Obsession:** pronounced preoccupation with some idea, often associated with anxiety.

47. **Oedipus complex:** Freud's idea that normal development involves developing (and eventually suppressing) sexual impulses toward the opposite-sex parent while viewing the other parent as a threat.
48. **Pathological:** related to disease.
49. **Personality:** an individual's characteristic pattern of thought and behavior.
50. **Phobia:** an intense, irrational fear.
51. **Preconscious:** the part of the mind that exists between the unconscious and the conscious. Memories the mind does not suppress exist here before they are called into consciousness.
52. **Prefrontal cortex:** an area of the brain associated with higher cognitive functions.
53. **Procedural memory:** memory of how to engage in tasks; often a person does not consciously recall this memory.
54. **Projection:** a Freudian ego defense mechanism by which an individual attributes his or her own uncomfortable thoughts to someone else.

55. **Psyche:** the mind.
56. **Psychiatry:** a branch of medicine involved in treating psychological disorders.
57. **Psychic:** in this context, concerning mental processes.
58. **Psychoanalysis:** a school of thought emphasizing the role the unconscious mind plays in conscious behavior; also a therapeutic approach that employs various techniques to try to access the unconscious mind.
59. **Psychodynamic:** a school of thought that emphasizes that unconscious forces shape behavior.
60. **Psychology:** the scientific study of mental processes and behaviors.
61. **Psychoneurosis (also neurosis):** mental illness involving anxiety.
62. **Psychopathology:** the study of the diseases of the mind.
63. **Rationalism:** a philosophical perspective emphasizing reason and logic.

64. **Rationalization:** a Freudian ego defense mechanism in which an individual tries to justify his or her behavior by explaining it.
65. **Reflex arc:** a popular concept in Freud's time, according to which behavior can be considered a reflexive reaction to sensory input or sensation.
66. **Reflex circuits:** neural circuits involving sensory input and motor output.
67. **Representation:** in this context, refers to the way the mind presents latent thoughts in the manifest content of the dream.
68. **Repression:** a Freudian ego defense mechanism that involves pushing uncomfortable thoughts and memories out of conscious awareness.
69. **Secondary elaboration:** the cohesive narrative the unconscious mind provides for the dream's manifest content.
70. **Sensory input:** information received by sensory receptors that is then transmitted to the brain.
71. **Stage theory of psychosexual development:** Freud's ideas that the resolution of conflicts center on the location of an individual's erogenous zones at a given period in development. Resolving—or not resolving—these conflicts has long-term consequences for the individual's personality.
72. **Superego:** the part of the mind that contains one's sense of conscience.
73. **Theory of mind:** Freud's attempt to explain the thought processes that occur in other people.
74. **Unconscious mind:** the portion of the mind that falls outside conscious awareness.
75. **Victorian era:** a specific period of history (roughly marked by the reign of Great Britain's Queen Victoria) from the nineteenth into the early twentieth century. Most importantly here, Victorians had a moral preoccupation with sexuality.
76. **Wet dream:** a dream that involves ejaculation.

PEOPLE MENTIONED IN THE TEXT

1. **Alfred Adler (1870—1937)** was an Austrian physician and psychotherapist. A former follower of Freud, Adler went on to found individual psychology.

2. **Josef Breuer (1842—1925)** was an Austrian physician who mentored Freud and collaborated with him on *Studies on Hysteria*.

3. **A. A. Brill (1874—1948)** was an Austrian psychiatrist who was the first to translate *The Interpretation of Dreams* into English.

4. **Jean-Martin Charcot (1825—1893)** was a French neurologist who mentored Freud. Charcot is often considered the father of neurology.

5. **Charles Darwin (1809—1882)** was a British naturalist best known for outlining the principles of the theory of evolution.

6. **Albert Einstein (1879—1955)** was a German-born physicist best known for his theory of general relativity.

7. **Sándor Ferenczi (1873—1933)** was a Hungarian psychoanalyst who was one of Freud's associates in the latter stages of his life. He diverged from Freud's ideas that the psychoanalyst should be a passive participant in the process and advocated instead for a more active role.

8. **Johann Joseph Gassner (1727—1779)** was an Austrian healer and exorcist. Gassner's confrontation with Franz Anton Mesmer set the stage for modern psychiatry.

9. **Peter Gay (1923—2015)** was a German American historian who specialized in European culture and intellectual history.

10. **Queen Jocasta** is a figure from Greek mythology, the wife of King Laius and mother to King Oedipus.

11. **Ernest Jones (1879—1958)** was a British psychoanalyst. Jones was also Freud's biographer.

12. **Carl Jung (1875—1961)** was a Swiss psychiatrist who was one of Freud's most prominent associates. Jung went on to establish the analytical perspective in psychiatry.

13. **Eric Kandel (b. 1929)** is an Austrian American neuropsychiatrist who won

the Nobel Prize in 2000 for his research on the biological basis of memory.

14. **King Laius** is a figure from Greek mythology, the father of King Oedipus.

15. **Karl Marx (1818—1883)** was a German philosopher best known for his books *The Communist Manifesto* and *Das Kapital*.

16. **Franz Anton Mesmer (1734—1815)** was a German physician who is best known for the concept of animal magnetism.

17. **Theodor Meynert (1833—1892)** was a German Austrian neuropathologist, who supervised Freud's early work in theVienna hospital's psychiatric ward.

18. **Anna O.** was the pseudonym given to Josef Breuer's most famous patient. Her case study served as an impetus for the book he cowrote with Freud.

19. **King Oedipus** is a figure from Greek mythology. He unwittingly murdered his father, King Laius, and married his mother, Queen Jocasta.

WORKS CITED

1. Bassiri, Nima. "Freud and the Matter of the Brain: On the Rearrangements of Neuropsychoanalysis." *Critical Inquiry* 40, no. 1 (2013): 83–108.
2. Cherry, Kendra. "Books by Sigmund Freud: Freud's Most Famous and Influential Books." *About.com.* http://psychology.about.com/od/sigmundfreud/tp/books-by-sigmund-freud.htm.
3. "Sigmund Freud Photobiography: Freud and Jung." *About.com.* http://psychology.about.com/od/sigmundfreud/ig/Sigmund-Freud-Photobiography/Freud-and-Jung.htm.
4. Eidenberg, David. "Freud and Jung: A 'Psychoanalysis' in Letters." *Psychological Perspectives* 57 (2014): 7–24.
5. Ellenberger, Henri F. *The Discovery of the Unconscious: The History and Evolution of Dynamic Psychiatry.* New York: Basic Books, 2008.
6. Frank, George. "A Response to 'The Relevance of Sigmund Freud for the 21st Century." *Psychoanalytic Psychology* 25, no. 2 (2008): 375–379.
7. Freud, Sigmund. *The Interpretation of Dreams.* Translated by A. A. Brill. Introduction and Notes by Daniel T. O'Hara and Gina Masucci MacKenzie. New York: Barnes & Noble Books, 2005.
8. "Freud's book, 'The Interpretation of Dreams' released 1900." *PBS.org.* Accessed February 17, 2016. http://www.pbs.org/wgbh/aso/databank/entries/dh00fr.html.
9. Freud, Sigmund, and Josef Breuer. *Studies on Hysteria.* Translated by James Strachey. London: Hogarth Press, 1955.
10. Gonzalez-Torres, Miguel Angel. "Psychoanalysis and Neuroscience. Friends or Enemies?" *International Forum of Psychoanalysis* 22, no. 1 (2013): 35–42.
11. "Historical Context for the Writings of Sigmund Freud." *Columbia College: The Core Curriculum.* http://www.college.columbia.edu/core/content/writings-sigmund-freud/context.
12. Jones, Ernest. "Freud and His Achievements." *The British Medical Journal* 1, no. 4974 (1956): 997–1000.
13. Jung, C. G. "Sigmund Freud in His Historical Setting." *Journal of Personality* 1, no. 1 (1932): 48–55.
14. Kandel, Eric R. "Biology and the Future of Psychoanalysis: A New Intellectual Framework for Psychiatry Revisited." *The American Journal of Psychiatry* 156,

no. 4 (1999): 505–524.

15. Kaunders, Anthony D. "Truth, Truthfulness, and Psychoanalysis: The Reception of Freud in Wilhelmine Germany." *German History* 31, no. 1 (2013): 1–22.

16. Kitcher, Patricia. *Freud's Dream: A Complete Interdisciplinary Science of Mind.* Cambridge, MA: MIT Press, 1992.

17. Libbrecht, K., and J. Quackelbeen. "On the Early History of Male Hysteria and Psychic Trauma. Charcot's Influence on Freudian Thought." *Journal of the History of the Behavioral Sciences* 31, no. 4 (1995): 370–384.

18. McHugh, Paul R. "The Death of Freud and the Rebirth of Psychiatry." *TheWeeklyStandard.com*. July 17, 2000. Accessed February 22, 2016. http://www.weeklystandard.com/article/12226.

19. McLeod, Saul. "Psychodynamic Approach." *SimplyPsychology*. 2007. Accessed February 22, 2016. http://www.simplypsychology.org/psychodynamic.html.

20. Muller, John P. "A Re-Reading of *Studies on Hysteria*: The Freud-Breuer Break Revisited." *Psychoanalytic Psychology* 9, no. 2 (1992): 129–156.

21. "Personality," in *Psychology*, 369–410 (Houston, TX: OpenStax College, 2014), 375.

22. Reppen, Joseph. "The Relevance of Sigmund Freud for the 21st Century." *Psychoanalytic Psychology* 23, no. 2 (2006): 215–216.

23. Ricoeur, Paul. "Sigmund Freud." In Karl Simms, *Paul Ricoeur: Routledge Critical Thinkers*, 46. Abingdon, UK: Taylor & Francis, 2002.

24. Solomon, Hester McFarland. "Freud and Jung: An Incomplete Encounter." *Journal of Analytical Psychology* 48, no. 5 (2003): 553–569.

25. Stengel, E. "Freud's Impact on Psychiatry." *The British Medical Journal* 1, no. 4974 (1956): 1000–1003.

26. Strenger, Carlo. "Can Psychoanalysis Reclaim the Public Sphere?" *Psychoanalytic Psychology* 32, no. 2 (2015): 239–306.

27. Thorne, B. Michael, and Tracy B. Henley. *Connections in the History and Systems of Psychology*. Boston, MA: Houghton Mifflin Company, 2005.

28. Wallerstein, Robert S. "The Relevance of Freud's Psychoanalysis in the 21st Century: Its Science and Its Research." *Psychoanalytic Psychology* 23, no. 2 (2006): 302–326.

29. Webster, Richard. "Freud, Charcot, and Hysteria: Lost in the Labyrinth." *RichardWebster.net*. Accessed February 22, 2016. http://www.richardwebster.net/freudandcharcot.html.
30. "The Well-Documented Friendship of Carl Jung and Sigmund Freud." *Historacle.org*. Accessed January 27, 2016. http://historacle.org/freud_jung.html.

原书作者简介

西格蒙德·弗洛伊德是"精神分析学之父"。他于 1856 年出生在奥地利的首都维也纳。他就读维也纳大学医学院，毕业后在家乡开设私人诊所。他与医生约瑟夫·布洛伊尔合作治疗神经症，后来写成《癔症研究》一书。弗洛伊德在此书基础上——并结合他对个案病人的治疗和对他本人梦境的分析——完成了《梦的解析》，构想并阐述了他的精神分析学说。这本著作不仅让他一举成名，而且在精神疾病治疗领域带来了一场革命。1938 年，纳粹德国吞并奥地利之前，弗洛伊德（他是犹太人）举家逃亡英国。翌年，他在伦敦逝世，享年 83 岁。

本书作者简介

威廉·詹金斯，密歇根大学心理学博士，乔治亚洲莫瑟尔大学心理学系主任。

世界名著中的批判性思维

世界思想宝库钥匙丛书致力于深入浅出地阐释全世界著名思想家的观点，不论是谁、在何处都能了解到，从而推进批判性思维发展。

世界思想宝库钥匙丛书与世界顶尖大学的一流学者合作，为一系列学科中最有影响的著作推出新的分析文本，介绍其观点和影响。在这一不断扩展的系列中，每种选入的著作都代表了历经时间考验的思想典范。通过为这些著作提供必要背景、揭示原作者的学术渊源以及说明这些著作所产生的影响，本系列图书希望让读者以新视角看待这些划时代的经典之作。读者应学会思考、运用并挑战这些著作中的观点，而不是简单接受它们。

ABOUT THE AUTHOR OF THE ORIGINAL WORK

Sigmund Freud, "the father of psychoanalysis," was born in Vienna, Austria, in 1856. He studied medicine at the University of Vienna before opening a private practice in his hometown. His work with physician Josef Breuer on treating nervous disorders led to a book, *Studies on Hysteria*.

Freud built on that—as well as on his work with private patients and analysis of his own dreams—to formulate the psychological theories he laid out in *The Interpretation of Dreams*. The text made him a celebrity, but it also revolutionized the treatment of mental illness. Just before Nazi Germany annexed Austria in 1938, Freud (who was Jewish) fled to England with his family. He died in London the following year at the age of 83.

ABOUT THE AUTHORS OF THE ANALYSIS

Dr William Jenkins holds a PhD in psychology from the University of Michigan. He is currently co-chair of the Department of Psychology at Mercer University in Georgia.

ABOUT MACAT
GREAT WORKS FOR CRITICAL THINKING

Macat is focused on making the ideas of the world's great thinkers accessible and comprehensible to everybody, everywhere, in ways that promote the development of enhanced critical thinking skills.

It works with leading academics from the world's top universities to produce new analyses that focus on the ideas and the impact of the most influential works ever written across a wide variety of academic disciplines. Each of the works that sit at the heart of its growing library is an enduring example of great thinking. But by setting them in context—and looking at the influences that shaped their authors, as well as the responses they provoked—Macat encourages readers to look at these classics and game-changers with fresh eyes. Readers learn to think, engage and challenge their ideas, rather than simply accepting them.

批判性思维与《梦的解析》

批判性思维的首要技巧：阐释

批判性思维的次要技巧：分析

西格蒙德·弗洛伊德 1899 年出版的《梦的解析》无疑是关于阐释与分析技巧最著名的作品。尽管该书在 8 年内仅售出 600 册，但它最终成为经典名著，让弗洛伊德声名卓著，成为 19 世纪与 20 世纪最重要的知识分子之一。

正如弗洛伊德的精神分析理论所展现的那样，在批判性思维中，阐释旨在理解材料的内涵，探究事物的意义；而分析则是从被阐释的材料中梳理出隐含的理据与构想。

《梦的解析》是对材料进行睿智阐释并作出有效分析的经典案例。弗洛伊德理论的出发点来自以下构想，即梦是无意识心理解决各种冲突的有意义的活动。因此，在弗洛伊德看来，梦含有被改变的、隐匿的形式线索，有助于认识我们最深层的无意识冲动和欲望。阐释者必须对每一条线索作出具体分析，然后解读其真实含义。尽管弗洛伊德的理论经常遭到批评，但是他仍然是无可争议的阐释大师——连他的批评者们也认为他是十分睿智的阐释者。

CRITICAL THINKING AND *THE INTERPRETATION OF DREAMS*

- Primary critical thinking skill: INTERPRETATION
- Secondary critical thinking skill: ANALYSIS

There is arguably no more famous book about the arts of interpretation and analysis than Sigmund Freud's 1899 *The Interpretation of Dreams*. Though the original edition of just 600 copies took eight years to sell out, the book eventually became a classic text that helped cement Freud's reputation as one of the most significant intellectual figures of the 19th and 20th centuries.

In critical thinking, just as in Freud's psychoanalytical theories, interpretation is all about understanding the meaning of evidence, and tracing the significance of things. Analysis can then be brought in to tease out the implicit reasons and assumptions that lie underneath the interpreted evidence.

The Interpretation of Dreams is a masterclass in building telling analyses from ingenious interpretation of evidence. Freud worked from the assumption that all dreams were significant attempts by the unconscious to resolve conflicts. As a result, he argued, they contain in altered and disguised forms clues to our deepest unconscious urges and desires. Each must be taken on its own terms to tease out what they really mean. Though Freud's theories have often been criticized, he remains the undisputed master of interpretation—with his critics suggesting that he was, if anything, too ingenious for his own good.

世界思想宝库钥匙丛书简介

世界思想宝库钥匙丛书致力于为一系列在各领域产生重大影响的人文社科类经典著作提供独特的学术探讨。每一本读物都不仅仅是原经典著作的内容摘要，而是介绍并深入研究原经典著作的学术渊源、主要观点和历史影响。这一丛书的目的是提供一套学习资料，以促进读者掌握批判性思维，从而更全面、深刻地去理解重要思想。

每一本读物分为3个部分：学术渊源、学术思想和学术影响，每个部分下有4个小节。这些章节旨在从各个方面研究原经典著作及其反响。

由于独特的体例，每一本读物不但易于阅读，而且另有一项优点：所有读物的编排体例相同，读者在进行某个知识层面的调查或研究时可交叉参阅多本该丛书中的相关读物，从而开启跨领域研究的路径。

为了方便阅读，每本读物最后还列出了术语表和人名表（在书中则以星号*标记），此外还有参考文献。

世界思想宝库钥匙丛书与剑桥大学合作，理清了批判性思维的要点，即如何通过6种技能来进行有效思考。其中3种技能让我们能够理解问题，另3种技能让我们有能力解决问题。这6种技能合称为"批判性思维PACIER模式"，它们是：

分析：了解如何建立一个观点；
评估：研究一个观点的优点和缺点；
阐释：对意义所产生的问题加以理解；
创造性思维：提出新的见解，发现新的联系；
解决问题：提出切实有效的解决办法；
理性化思维：创建有说服力的观点。

了解更多信息，请浏览www.macat.com。

THE MACAT LIBRARY

The Macat Library is a series of unique academic explorations of seminal works in the humanities and social sciences — books and papers that have had a significant and widely recognised impact on their disciplines. It has been created to serve as much more than just a summary of what lies between the covers of a great book. It illuminates and explores the influences on, ideas of, and impact of that book. Our goal is to offer a learning resource that encourages critical thinking and fosters a better, deeper understanding of important ideas.

Each publication is divided into three Sections: Influences, Ideas, and Impact. Each Section has four Modules. These explore every important facet of the work, and the responses to it.

This Section-Module structure makes a Macat Library book easy to use, but it has another important feature. Because each Macat book is written to the same format, it is possible (and encouraged!) to cross-reference multiple Macat books along the same lines of inquiry or research. This allows the reader to open up interesting interdisciplinary pathways.

To further aid your reading, lists of glossary terms and people mentioned are included at the end of this book (these are indicated by an asterisk [*] throughout) — as well as a list of works cited.

Macat has worked with the University of Cambridge to identify the elements of critical thinking and understand the ways in which six different skills combine to enable effective thinking.

Three allow us to fully understand a problem; three more give us the tools to solve it. Together, these six skills make up the PACIER model of critical thinking. They are:

ANALYSIS — understanding how an argument is built
EVALUATION — exploring the strengths and weaknesses of an argument
INTERPRETATION — understanding issues of meaning
CREATIVE THINKING — coming up with new ideas and fresh connections
PROBLEM-SOLVING — producing strong solutions
REASONING — creating strong arguments

To find out more, visit WWW.MACAT.COM.

"世界思想宝库钥匙丛书提供了独一无二的跨学科学习和研究工具。它介绍那些革新了各自学科研究的经典著作,还邀请全世界一流专家和教育机构进行严谨的分析,为每位读者打开世界顶级教育的大门。"

—— 安德烈亚斯·施莱歇尔,
经济合作与发展组织教育与技能司司长

"世界思想宝库钥匙丛书直面大学教育的巨大挑战……他们组建了一支精干而活跃的学者队伍,来推出在研究广度上颇具新意的教学材料。"

—— 布罗尔斯教授、勋爵,剑桥大学前校长

"世界思想宝库钥匙丛书的愿景令人赞叹。它通过分析和阐释那些曾深刻影响人类思想以及社会、经济发展的经典文本,提供了新的学习方法。它推动批判性思维,这对于任何社会和经济体来说都是至关重要的。这就是未来的学习方法。"

—— 查尔斯·克拉克阁下,英国前教育大臣

"对于那些影响了各自领域的著作,世界思想宝库钥匙丛书能让人们立即了解到围绕那些著作展开的评论性言论,这让该系列图书成为在这些领域从事研究的师生们不可或缺的资源。"

—— 威廉·特朗佐教授,加利福尼亚大学圣地亚哥分校

"Macat offers an amazing first-of-its-kind tool for interdisciplinary learning and research. Its focus on works that transformed their disciplines and its rigorous approach, drawing on the world's leading experts and educational institutions, opens up a world-class education to anyone."

—— Andreas Schleicher, Director for Education and Skills, Organisation for Economic Co-operation and Development

"Macat is taking on some of the major challenges in university education... They have drawn together a strong team of active academics who are producing teaching materials that are novel in the breadth of their approach."

—— Prof Lord Broers, former Vice-Chancellor of the University of Cambridge

"The Macat vision is exceptionally exciting. It focuses upon new modes of learning which analyse and explain seminal texts which have profoundly influenced world thinking and so social and economic development. It promotes the kind of critical thinking which is essential for any society and economy. This is the learning of the future."

—— Rt Hon Charles Clarke, former UK Secretary of State for Education

"The Macat analyses provide immediate access to the critical conversation surrounding the books that have shaped their respective discipline, which will make them an invaluable resource to all of those, students and teachers, working in the field."

—— Prof William Tronzo, University of California at San Diego

The Macat Libary
世界思想宝库钥匙丛书

TITLE	中文书名	类别
An Analysis of Arjun Appadurai's *Modernity at Large: Cultural Dimensions of Globalisation*	解析阿尔君·阿帕杜莱《消失的现代性：全球化的文化维度》	人类学
An Analysis of Claude Lévi-Strauss's *Structural Anthropology*	解析克劳德·列维-斯特劳斯《结构人类学》	人类学
An Analysis of Marcel Mauss's *The Gift*	解析马塞尔·莫斯《礼物》	人类学
An Analysis of Jared M. Diamond's *Guns, Germs, and Steel: The Fate of Human Societies*	解析贾雷德·戴蒙德《枪炮、病菌与钢铁：人类社会的命运》	人类学
An Analysis of Clifford Geertz's *The Interpretation of Cultures*	解析克利福德·格尔茨《文化的解释》	人类学
An Analysis of Philippe Ariès's *Centuries of Childhood: A Social History of Family Life*	解析菲力浦·阿利埃斯《儿童的世纪：旧制度下的儿童和家庭生活》	人类学
An Analysis of W. Chan Kim & Renée Mauborgne's *Blue Ocean Strategy*	解析金伟灿/勒妮·莫博涅《蓝海战略》	商业
An Analysis of John P. Kotter's *Leading Change*	解析约翰·P.科特《领导变革》	商业
An Analysis of Michael E. Porter's *Competitive Strategy: Creating and Sustaining Superior Performance*	解析迈克尔·E.波特《竞争战略：分析产业和竞争对手的技术》	商业
An Analysis of Jean Lave & Etienne Wenger's *Situated Learning: Legitimate Peripheral Participation*	解析琼·莱夫/艾蒂纳·温格《情境学习：合法的边缘性参与》	商业
An Analysis of Douglas McGregor's *The Human Side of Enterprise*	解析道格拉斯·麦格雷戈《企业的人性面》	商业
An Analysis of Milton Friedman's *Capitalism and Freedom*	解析米尔顿·弗里德曼《资本主义与自由》	商业
An Analysis of Ludwig von Mises's *The Theory of Money and Credit*	解析路德维希·冯·米塞斯《货币和信用理论》	经济学
An Analysis of Adam Smith's *The Wealth of Nations*	解析亚当·斯密《国富论》	经济学
An Analysis of Thomas Piketty's *Capital in the Twenty-First Century*	解析托马斯·皮凯蒂《21世纪资本论》	经济学
An Analysis of Nassim Nicholas Taleb's *The Black Swan: The Impact of the Highly Improbable*	解析纳西姆·尼古拉斯·塔勒布《黑天鹅：如何应对不可预知的未来》	经济学
An Analysis of Ha-Joon Chang's *Kicking Away the Ladder*	解析张夏准《富国陷阱：发达国家为何踢开梯子》	经济学
An Analysis of Thomas Robert Malthus's *An Essay on the Principle of Population*	解析托马斯·马尔萨斯《人口论》	经济学

An Analysis of John Maynard Keynes's *The General Theory of Employment, Interest and Money*	解析约翰·梅纳德·凯恩斯《就业、利息和货币通论》	经济学
An Analysis of Milton Friedman's *The Role of Monetary Policy*	解析米尔顿·弗里德曼《货币政策的作用》	经济学
An Analysis of Burton G. Malkiel's *A Random Walk Down Wall Street*	解析伯顿·G.马尔基尔《漫步华尔街》	经济学
An Analysis of Friedrich A. Hayek's *The Road to Serfdom*	解析弗里德里希·A.哈耶克《通往奴役之路》	经济学
An Analysis of Charles P. Kindleberger's *Manias, Panics, and Crashes: A History of Financial Crises*	解析查尔斯·P.金德尔伯格《疯狂、惊恐和崩溃：金融危机史》	经济学
An Analysis of Amartya Sen's *Development as Freedom*	解析阿玛蒂亚·森《以自由看待发展》	经济学
An Analysis of Rachel Carson's *Silent Spring*	解析蕾切尔·卡森《寂静的春天》	地理学
An Analysis of Charles Darwin's *On the Origin of Species: by Means of Natural Selection, or The Preservation of Favoured Races in the Struggle for Life*	解析查尔斯·达尔文《物种起源》	地理学
An Analysis of World Commission on Environment and Development's *The Brundtland Report, Our Common Future*	解析世界环境与发展委员会《布伦特兰报告：我们共同的未来》	地理学
An Analysis of James E. Lovelock's *Gaia: A New Look at Life on Earth*	解析詹姆斯·E.拉伍洛克《盖娅：地球生命的新视野》	地理学
An Analysis of Paul Kennedy's *The Rise and Fall of the Great Powers: Economic Change and Military Conflict from 1500—2000*	解析保罗·肯尼迪《大国的兴衰：1500—2000年的经济变革与军事冲突》	历史
An Analysis of Janet L. Abu-Lughod's *Before European Hegemony: The World System A. D. 1250—1350*	解析珍妮特·L.阿布-卢格霍德《欧洲霸权之前：1250—1350年的世界体系》	历史
An Analysis of Alfred W. Crosby's *The Columbian Exchange: Biological and Cultural Consequences of 1492*	解析艾尔弗雷德·W.克罗斯比《哥伦布大交换：1492年以后的生物影响和文化冲击》	历史
An Analysis of Tony Judt's *Postwar: A History of Europe since 1945*	解析托尼·贾德《战后欧洲史》	历史
An Analysis of Richard J. Evans's *In Defence of History*	解析理查德·J.艾文斯《捍卫历史》	历史
An Analysis of Eric Hobsbawm's *The Age of Revolution: Europe 1789—1848*	解析艾瑞克·霍布斯鲍姆《革命的年代：欧洲1789—1848年》	历史

An Analysis of Roland Barthes's *Mythologies*	解析罗兰·巴特《神话学》	文学与批判理论
An Analysis of Simon de Beauvoir's *The Second Sex*	解析西蒙娜·德·波伏娃《第二性》	文学与批判理论
An Analysis of Edward W. Said's *Orientalism*	解析爱德华·W. 萨义德《东方主义》	文学与批判理论
An Analysis of Virginia Woolf's *A Room of One's Own*	解析弗吉尼亚·伍尔芙《一间自己的房间》	文学与批判理论
An Analysis of Judith Butler's *Gender Trouble*	解析朱迪斯·巴特勒《性别麻烦》	文学与批判理论
An Analysis of Ferdinand de Saussure's *Course in General Linguistics*	解析费尔迪南·德·索绪尔《普通语言学教程》	文学与批判理论
An Analysis of Susan Sontag's *On Photography*	解析苏珊·桑塔格《论摄影》	文学与批判理论
An Analysis of Walter Benjamin's *The Work of Art in the Age of Mechanical Reproduction*	解析瓦尔特·本雅明《机械复制时代的艺术作品》	文学与批判理论
An Analysis of W.E.B. Du Bois's *The Souls of Black Folk*	解析W.E.B. 杜博伊斯《黑人的灵魂》	文学与批判理论
An Analysis of Plato's *The Republic*	解析柏拉图《理想国》	哲学
An Analysis of Plato's *Symposium*	解析柏拉图《会饮篇》	哲学
An Analysis of Aristotle's *Metaphysics*	解析亚里士多德《形而上学》	哲学
An Analysis of Aristotle's *Nicomachean Ethics*	解析亚里士多德《尼各马可伦理学》	哲学
An Analysis of Immanuel Kant's *Critique of Pure Reason*	解析伊曼努尔·康德《纯粹理性批判》	哲学
An Analysis of Ludwig Wittgenstein's *Philosophical Investigations*	解析路德维希·维特根斯坦《哲学研究》	哲学
An Analysis of G.W.F. Hegel's *Phenomenology of Spirit*	解析G.W.F. 黑格尔《精神现象学》	哲学
An Analysis of Baruch Spinoza's *Ethics*	解析巴鲁赫·斯宾诺莎《伦理学》	哲学
An Analysis of Hannah Arendt's *The Human Condition*	解析汉娜·阿伦特《人的境况》	哲学
An Analysis of G.E.M. Anscombe's *Modern Moral Philosophy*	解析G.E.M. 安斯康姆《现代道德哲学》	哲学
An Analysis of David Hume's *An Enquiry Concerning Human Understanding*	解析大卫·休谟《人类理解研究》	哲学

An Analysis of Søren Kierkegaard's *Fear and Trembling*	解析索伦·克尔凯郭尔《恐惧与战栗》	哲学
An Analysis of René Descartes's *Meditations on First Philosophy*	解析勒内·笛卡尔《第一哲学沉思录》	哲学
An Analysis of Friedrich Nietzsche's *On the Genealogy of Morality*	解析弗里德里希·尼采《论道德的谱系》	哲学
An Analysis of Gilbert Ryle's *The Concept of Mind*	解析吉尔伯特·赖尔《心的概念》	哲学
An Analysis of Thomas Kuhn's *The Structure of Scientific Revolutions*	解析托马斯·库恩《科学革命的结构》	哲学
An Analysis of John Stuart Mill's *Utilitarianism*	解析约翰·斯图亚特·穆勒《功利主义》	哲学
An Analysis of Aristotle's *Politics*	解析亚里士多德《政治学》	政治学
An Analysis of Niccolò Machiavelli's *The Prince*	解析尼科洛·马基雅维利《君主论》	政治学
An Analysis of Karl Marx's *Capital*	解析卡尔·马克思《资本论》	政治学
An Analysis of Benedict Anderson's *Imagined Communities*	解析本尼迪克特·安德森《想象的共同体》	政治学
An Analysis of Samuel P. Huntington's *The Clash of Civilizations and the Remaking of World Order*	解析塞缪尔·P.亨廷顿《文明的冲突与世界秩序重建》	政治学
An Analysis of Alexis de Tocqueville's *Democracy in America*	解析阿列克西·德·托克维尔《论美国的民主》	政治学
An Analysis of J. A. Hobson's *Imperialism: A Study*	解析约·阿·霍布森《帝国主义》	政治学
An Analysis of Thomas Paine's *Common Sense*	解析托马斯·潘恩《常识》	政治学
An Analysis of John Rawls's *A Theory of Justice*	解析约翰·罗尔斯《正义论》	政治学
An Analysis of Francis Fukuyama's *The End of History and the Last Man*	解析弗朗西斯·福山《历史的终结与最后的人》	政治学
An Analysis of John Locke's *Two Treatises of Government*	解析约翰·洛克《政府论》	政治学
An Analysis of Sun Tzu's *The Art of War*	解析孙武《孙子兵法》	政治学
An Analysis of Henry Kissinger's *World Order: Reflections on the Character of Nations and the Course of History*	解析亨利·基辛格《世界秩序》	政治学
An Analysis of Jean-Jacques Rousseau's *The Social Contract*	解析让-雅克·卢梭《社会契约论》	政治学

An Analysis of Odd Arne Westad's *The Global Cold War: Third World Interventions and the Making of Our Times*	解析文安立《全球冷战：美苏对第三世界的干涉与当代世界的形成》	政治学
An Analysis of Sigmund Freud's *The Interpretation of Dreams*	解析西格蒙德·弗洛伊德《梦的解析》	心理学
An Analysis of William James' *The Principles of Psychology*	解析威廉·詹姆斯《心理学原理》	心理学
An Analysis of Philip Zimbardo's *The Lucifer Effect*	解析菲利普·津巴多《路西法效应》	心理学
An Analysis of Leon Festinger's *A Theory of Cognitive Dissonance*	解析利昂·费斯汀格《认知失调论》	心理学
An Analysis of Richard H. Thaler & Cass R. Sunstein's *Nudge: Improving Decisions about Health, Wealth, and Happiness*	解析理查德·H.泰勒/卡斯·R.桑斯坦《助推：如何做出有关健康、财富和幸福的更优决策》	心理学
An Analysis of Gordon Allport's *The Nature of Prejudice*	解析高尔登·奥尔波特《偏见的本质》	心理学
An Analysis of Steven Pinker's *The Better Angels of Our Nature: Why Violence Has Declined*	解析斯蒂芬·平克《人性中的善良天使：暴力为什么会减少》	心理学
An Analysis of Stanley Milgram's *Obedience to Authority*	解析斯坦利·米尔格拉姆《对权威的服从》	心理学
An Analysis of Betty Friedan's *The Feminine Mystique*	解析贝蒂·弗里丹《女性的奥秘》	心理学
An Analysis of David Riesman's *The Lonely Crowd: A Study of the Changing American Character*	解析大卫·理斯曼《孤独的人群：美国人社会性格演变之研究》	社会学
An Analysis of Franz Boas's *Race, Language and Culture*	解析弗朗兹·博厄斯《种族、语言与文化》	社会学
An Analysis of Pierre Bourdieu's *Outline of a Theory of Practice*	解析皮埃尔·布尔迪厄《实践理论大纲》	社会学
An Analysis of Max Weber's *The Protestant Ethic and the Spirit of Capitalism*	解析马克斯·韦伯《新教伦理与资本主义精神》	社会学
An Analysis of Jane Jacobs's *The Death and Life of Great American Cities*	解析简·雅各布斯《美国大城市的死与生》	社会学
An Analysis of C. Wright Mills's *The Sociological Imagination*	解析C.赖特·米尔斯《社会学的想象力》	社会学
An Analysis of Robert E. Lucas Jr.'s *Why doesn't Capital Flow from Rich to Poor Countries?*	解析小罗伯特·E.卢卡斯《为何资本不从富国流向穷国？》	社会学

An Analysis of Émile Durkheim's *On Suicide*	解析埃米尔·迪尔凯姆《自杀论》	社会学
An Analysis of Eric Hoffer's *The True Believer: Thoughts on the Nature of Mass Movements*	解析埃里克·霍弗《狂热分子：群众运动圣经》	社会学
An Analysis of Jared M. Diamond's *Collapse: How Societies Choose to Fail or Survive*	解析贾雷德·M. 戴蒙德《大崩溃：社会如何选择兴亡》	社会学
An Analysis of Michel Foucault's *The History of Sexuality Vol. 1: The Will to Knowledge*	解析米歇尔·福柯《性史（第一卷）：求知意志》	社会学
An Analysis of Michel Foucault's *Discipline and Punish*	解析米歇尔·福柯《规训与惩罚》	社会学
An Analysis of Richard Dawkins's *The Selfish Gene*	解析理查德·道金斯《自私的基因》	社会学
An Analysis of Antonio Gramsci's *Prison Notebooks*	解析安东尼奥·葛兰西《狱中札记》	社会学
An Analysis of Augustine's *Confessions*	解析奥古斯丁《忏悔录》	神学
An Analysis of C. S. Lewis's *The Abolition of Man*	解析C.S. 路易斯《人之废》	神学

图书在版编目（CIP）数据

解析西格蒙德·弗洛伊德《梦的解析》/ 威廉·詹金斯（William J. Jenkins）著；张和龙译. —上海：上海外语教育出版社，2019
（世界思想宝库钥匙丛书）
ISBN 978-7-5446-5774-7

Ⅰ.①解… Ⅱ.①威…②张… Ⅲ.①弗洛伊德—精神分析—研究 Ⅳ.①B84-065

中国版本图书馆CIP数据核字（2019）第044904号

This Chinese-English bilingual edition of *An Analysis of Sigmund Freud's* The Interpretation of Dreams is published by arrangement with Macat International Limited.
Licensed for sale throughout the world.
本书汉英双语版由Macat国际有限公司授权上海外语教育出版社有限公司出版。
供在全世界范围内发行、销售。

图字：09 – 2018 – 549

出版发行：上海外语教育出版社
　　　　　（上海外国语大学内）　邮编：200083
电　　话：021-65425300（总机）
电子邮箱：bookinfo@sflep.com.cn
网　　址：http://www.sflep.com
责任编辑：杨莹雪

印　　刷：上海信老印刷厂
开　　本：890×1240　1/32　印张 5.5　字数 140千字
版　　次：2019年8月第1版　2019年8月第1次印刷
印　　数：2 100 册

书　　号：ISBN 978-7-5446-5774-7 / B
定　　价：30.00 元
　　　本版图书如有印装质量问题，可向本社调换
　　　质量服务热线：4008-213-263　电子邮箱：editorial@sflep.com